The Life of Professor Robert Hugh Pritchard

The Rise of Genetics at Leicester

The Life of Professor Robert Hugh Pritchard

The Rise of Genetics at Leicester

Editor

Arieh Zaritsky

Ben-Gurion University of the Negev, Israel

World Scientific

NEW JERSEY • LONDON • SINGAPORE • BEIJING • SHANGHAI • HONG KONG • TAIPEI • CHENNAI • TOKYO

Published by

World Scientific Publishing Co. Pte. Ltd.
5 Toh Tuck Link, Singapore 596224
USA office: 27 Warren Street, Suite 401-402, Hackensack, NJ 07601
UK office: 57 Shelton Street, Covent Garden, London WC2H 9HE

British Library Cataloguing-in-Publication Data
A catalogue record for this book is available from the British Library.

THE LIFE OF PROFESSOR ROBERT HUGH PRITCHARD
The Rise of Genetics at Leicester

ISBN 978-981-3203-73-0
ISBN 978-981-3203-74-7 (pbk)

Printed in Singapore by Mainland Press Pte Ltd.

The Life of Professor Robert Hugh Pritchard (Bob): The Rise of Genetics at Leicester

ca 2000

Born: 25th January 1930, Wandsworth, London.

Married twice, secondly to the artist, writer and teacher Suzi Pritchard (deceased).

Two sons: John, Simon (both deceased) and one daughter: Naomi Matthews.

Founder, Professor (1964 to 1989, then Emeritus) and Head (1964 to 1983): Department of Genetics, University of Leicester.

Liberal Democrat: City and County Councillor (1989 to 2002).

Died: 12th April 2015, Leicester.

EMBO Workshop: "Incompatibilty and Control of Replication in Plasmids", Sønderborg, Denmark, 1978, organised by Bob Pritchard and Kurt Nordström.

Partial List of participants, projected linearly from *left-to-right* (and see a variation in color at page 8):

Kenichi Matsubara, Ken Timmis, John Collins, Peter Meacock, Jorge Crosa, X, Dick Novick, X, Dean Taylor, Marilyn Nugent, Kaspar von Meyenburg, Bob Rownd, Martine Couturier, X, Jim Wechsler, Lucien Caro, June Scott, X, X, Knud Rasmussen, Willie and son (David) Donachie, X, Fleming Hansen, X, X, Millie Masters, John Nijkamp, Bruce Kline, X, X, Jenny Broome-Smith, Hilary Richards, Michael Yarmolinsky, X, X, Dietmar Blohm, Soto Hiraga, X, Kurt Nordström, Bob Pritchard, Werner Maas, Walter Messer, Petter Gustafsson, X.

Preface

The late Professor Robert (Bob) Pritchard was a pivotal figure in the development of Microbial Genetics and Physiology research in the UK during the second half of the 20th century, a period of exciting discoveries that led to the new era of "Molecular Biology". Pritchard was the founding Professor of the Department of Genetics at the University of Leicester, recognised today as one of the UK's foremost university Genetics departments that produces research of international quality and world-wide significance.

This volume comprises a series of articles written by several of his research collaborators, students and trainees as personal anecdotes and memories of their individual experiences and interactions with this eminent scientist. Amongst the contributors are several who came to him from countries and cultures outside of the UK who were able to take his liberal humanist philosophy back to their own homes.

The Department of Genetics, University of Leicester, UK, Summer 1971

Sitting: Siddhu (Sid) Singh, Jenny Barron (Foxon), Simon Hardy, Robert H Pritchard (Bob), Ned Holt, Ezra Yagil, Alan Wheals, Arieh Zaritsky, Brian Wilkins.

Middle: Val Darby, Sue Garrett, Jenny Dee, Robert Semeonoff, Terry Lymn, John Collins, Shamim Ahmad, Ifor R. Beacham, Horadigala G Nandadasa (Das), Clive (Clegg) Waldron.

Top: Susan Armitt (Grant), Judy Phipps, Bernard Senior, Hilda Statham, Tony Samson, Kathryn Beacham, Jim Mackley, Susan Hollom (Wilkins), Mick Chandler, Bill Grant, Clive Roberts, Mike Worsey, Aman Ansari, Russell Poulter, Pete Fantes.

(Photograph courtesy of Department of Genetics, University of Leicester.)

Robert Hugh Pritchard: Scientific Publications

Here we have attempted to make a comprehensive list of the scientific writings of R. H. Pritchard. Only the substantial and peer-reviewed publications are listed; abstracts of the many conference presentations that were prepared with his numerous trainees are not included.

Pritchard RH. (1953) Ascospores with diploid nuclei in *Aspergillus nidulans. Proc 9th Mt Cong Genet* (Abstract).

Pritchard RH, Pontecorvo G. (1953) The formation of ascospores with diploid nuclei in *Aspergillus nidulans. Microb Genet Bull* **7**: 18.

Pritchard RH. (1954) The relationship between a group of alleles in the *ad5* region of *Aspergillus nidulans. Heredity* **8**: 433 (Abstract).

Roper JA, Pritchard RH. (1955) Recovery of the complementary products of mitotic crossing over. *Nature* **175**: 639.

Pritchard RH. (1955) The linear arrangement of a series of alleles of *Aspergillus nidulans. Heredity* **9**: 343–371.

Pritchard RH. (1960) Localized negative interference and its bearing on models of gene recombination. *Genet Res* **1**: 1–24.

Pritchard RH. (1960) The bearing of recombination analysis at high resolution on genetic fine structure of *Aspergillus nidulans* and the mechanism of recombination in higher organisms. *Symp Soc Gen Microbiol* **10**: 155–180.

Mackintosh ME, Pritchard RH. (1963) The production and replica plating of microcolonies of *Aspergillus nidulans. Genet Res* **4**: 320–322.

Pritchard RH. (1963) The Relationship between conjugation, recombination and DNA synthesis in *Escherichia coli. Genetics Today: Proc Intern Cong Genet, 11th, The Hague*, pp. 55–78.

Pritchard RH, Lark KG. (1964) Induction of replication by thymine starvation at the chromosome origin in *Escherichia coli. J Mol Biol* **9**: 288–307.

Clowes RC, Moody EE, Pritchard RH. (1965) The elimination of extrachromosomal elements in thymineless strains of *Escherichia coli* K12. *Genet Res* **6**: 147–152.

Hollom S, Pritchard RH. (1965) Effect of inhibition of DNA synthesis on mating in *Escherichia coli* K12. *Genet Res* **6**: 479–483.

Kelly MS, Pritchard RH. (1965) Unstable linkage between genetic markers in transformation. *J Bacteriol* **89**: 1314–1321.

Pritchard RH. (1965) Structure and replication of the bacterial chromosome. *Brit Med Bull* **21**: 203–205.

Venema G, Pritchard RH, Venema-Schröder T. (1965) Fate of transforming deoxyribonucleic acid in *Bacillus subtilis*. *J Bacteriol* **89**: 1250–1255.

Venema G, Pritchard RH, Venema-Schröder T. (1965) Properties of newly introduced transforming deoxyribonucleic acid in *Bacillus subtilis*. *J Bacteriol* **90**: 343–346.

Pritchard RH. (1966) Review. Replication of the bacterial chromosome. *Proc R Soc Lond B Biol Sci* **164**: 258–266.

Pritchard RH. (1966) Contagious drug resistance in bacteria. *J Roy Coll Surgeons Edinburgh* **11**: 123–125.

Barth PT, Beacham IR, Ahmad SI, Pritchard RH. (1967) Properties of bacterial mutants defective in the catabolism of deoxynucleosides, *Biochem J* **106**: 36–37.

Pritchard RH. (1968) Genetics of *Aspergillus nidulans*. In: *Experiments in Microbial Genetics* (eds. RC Clowes and W Hayes), Blackwell Scientific Publications, Oxford and Edinburgh, pp. 159–183.

Ahmad SI, Barth PT, Pritchard RH. (1968) Properties of a mutant of *Escherichia coli* lacking purine nucleoside phosphorylase. *Biochim Biophys Acta* **161**: 581–583.

Barth PT, Beacham IR, Ahmad SI, Pritchard RH. (1968) The inducer of the deoxynucleoside phosphorylases and deoxyriboaldolase in *Escherichia coli*. *Biochim Biophys Acta* **161**: 554–557.

Beacham IR, Barth PT, Pritchard RH. (1968) Constitutivity of thymidine phosphorylase in deoxyriboaldolase negative strains: dependence on thymine requirement and concentration. *Biochim Biophys Acta* **166**: 589–592.

Beacham IR, Eisenstark A, Barth PT, Pritchard RH. (1968) Deoxynucleoside-sensitive mutants of *Salmonella typhimurium*. *Mol Gen Genet* **102**: 112–127.

Pritchard RH. Control of DNA synthesis in bacteria. (1968) *Heredity* **23**: 472–473 (Abstract).

Ahmad SI, Pritchard RH. (1969) A map of four genes specifying enzymes involved in catabolism of nucleosides and deoxynucleosides in *Escherichia coli*. *Mol Gen Genet* **104**: 351–359.

Pritchard RH, Barth PT, Collins J. (1969) Control of DNA synthesis in bacteria. *Microbial Growth. Symp Soc Gen Microbiol* **19**: 263–297.

Pritchard RH. (1969) Control of Replication of Genetic Material in Bacteria. Chapter 5 *in Ciba Foundation Symposium — Bacterial Episomes and Plasmids* (eds. GEW Wolstenholme and M O'Connor) John Wiley & Sons, Ltd., Chichester, UK, pp. 65–74.

Pritchard RH, Zaritsky A. (1970) Effect of thymine concentration on the replication velocity of DNA in a thymineless mutant of *Escherichia coli*. *Nature* **226**: 126–131.

Ahmad SI, Pritchard RH. (1971) A regulatory mutant affecting the synthesis of enzymes involved in the catabolism of nucleosides in *Escherichia coli*. *Mol Gen Genet* **111**: 77–83.

Beacham IR, Beacham K, Zaritsky A, Pritchard RH. (1971) Intracellular thymidine triphosphate concentrations in wild type and in thymine requiring mutants of *Escherichia coli* 15 and K12. *J Mol Biol* **60**: 75–86.

Zaritsky A, Pritchard RH. (1971) Replication time of the chromosome in thymineless mutants of *Escherichia coli*. *J Mol Biol* **60**: 65–74.

Beacham IR, Pritchard RH. (1971) The role of nucleoside phosphorylases in the degradation of deoxyribonucleosides by thymine-requiring mutants of *E. coli*. *Mol Gen Genet* **110**: 289–298.

Beacham IR, Yagil E, Beacham K, Pritchard RH. (1971) On the localisation of enzymes of deoxynucleoside catabolism in *Escherichia coli*. *FEBS Lett* **16**: 77–80.

Pritchard RH, Ahmad SI. (1971) Fluorouracil and the isolation of mutants lacking uridine phosphorylase in *Escherichia coli*: location of the gene. *Mol Gen Genet* **111**: 84–88.

Ahmad SI, Pritchard RH. (1972) Location of gene specifying cytosine deaminase in *Escherichia coli*. *Mol Gen Genet* **118**: 323–325.

Ahmad SI, Pritchard RH. (1973) An operator constitutive mutant affecting the synthesis of two enzymes involved in the catabolism of nucleosides in *Escherichia coli*. *Mol Gen Genet* **124**: 321–328.

Collins J, Pritchard RH. (1973) Relationship between chromosome replication and F'lac episome replication in *Escherichia coli*. *J Mol Biol* **78**: 143–155.

Zaritsky A, Pritchard RH. (1973) Changes in cell size and shape associated with changes in the replication time of the chromosome of *Escherichia coli*. *J Bacteriol* **114**: 824–837.

Pritchard RH. (1974) On the growth and form of a bacterial cell. *Phil Trans R Soc Lond B* **267**: 303–336.

Chandler MG, Pritchard RH. (1975) The effect of gene concentration and relative gene dosage on gene output in *Escherichia coli*. *Mol Gen Genet* **138**: 127–141.

Fantes PA, Grant WD, Pritchard RH, *et al.* (1975) The regulation of cell size and the control of mitosis. *J Theor Biol* **50**: 213–244.

Meacock PA, Pritchard RH. (1975) Relationship between chromosome replication and cell division in a thymineless mutant of *Escherichia coli* B/r. *J Bacteriol* **122**: 931–942.

Pritchard RH, Chandler MG, Collins J. (1975) Independence of F replication and chromosome replication in *Escherichia coli. Mol Gen Genet* **138**: 143–155.

Tresguerres EF, Nandadasa HG, Pritchard RH. (1975) Suppression of initiation-negative strains of *Escherichia coli* by integration of the sex factor F. *J Bacteriol* **121**: 554–561.

Diaz R, Pritchard RH. (1978) Cloning of replication origins from the *E. coli* K12 chromosome. *Nature* **275**: 561–564.

Meacock PA, Pritchard RH, Roberts EM. (1978) Effect of thymine concentration on cell shape in Thy$^-$ *Escherichia coli* B/r. *J Bacteriol* **133**: 320–328.

Pritchard RH, Meacock PA, Orr E. (1978) Diameter of cells of a thermosensitive *dnaA* mutant of *Escherichia coli* cultivated at intermediate temperatures. *J Bacteriol* **135**: 575–580.

Pritchard RH. (1978) Recombinant DNA is safe. *Nature* **273**: 696.

Pritchard RH. (1978) Control of DNA Replication in Bacteria. In: *DNA Synthesis: Present and Future* (eds. I Molineux and M Kohiyama), Volume 17 of the series NATO Advanced Study Institutes Series, Plenum Press, New York, pp. 1–26.

Orr E, Meacock PA, Pritchard RH. (1978) Genetic and physiological properties of an *Escherichia coli* strain carrying the *dnaA* mutation T46. In: *NATO Advanced Study Institute on DNA synthesis: present and future, Vol. 17e* (eds. IJ Molineux and M Kohiyama), New York: Plenum Press, pp. 85–100.

Diaz R, Barnsley P, Pritchard RH. (1979) Location and characterisation of a new replication origin in the *E. coli* K12 chromosome. *Mol Gen Genet* **175**: 151–157.

Orr E, Fairweather NF, Holland IB, Pritchard RH. (1979) Isolation and characterisation of a strain carrying a conditional lethal mutation in the *cou* gene of *Escherichia coli* K12. *Mol Gen Genet* **177**: 103–112.

Pritchard RH. (1979) Angels and devils of science. Review of "Recombinant DNA: The Untold Story" by J. Lear, 280 pp, Crown: New York, 1978. *Nature* **278**: 481–482. doi:10.1038/278481a0

Lycett GW, Orr E, Pritchard RH. (1980) Chloramphenicol releases a block in initiation of chromosome replication in a *dnaA* strain of *Escherichia coli* K12. *Mol Gen Genet* **178**: 329–336.

Pritchard RH. (1985) Control of chromosome replication in bacteria. *Basic Life Sci* **30**: 277–282.

Pritchard RH & Holland I.B (eds.). (1985) *Basic Cloning Techniques: A Manual of Experimental Procedures*, Blackwell Scientific Publications, Oxford.

Lycett GW, Pritchard RH. (1986) Functioning of the F-plasmid origin of replication in an *Escherichia coli* K12 Hfr strain during exponential growth. *Plasmid* **16**: 168–174.

Guzman EC, Jiménez-Sánchez A, Orr E, Pritchard RH. (1988) Heat stress in the presence of low RNA polymerase activity increases chromosome copy number of *Escherichia coli. Mol Gen Genet* **212**: 203–206.

Guzman EC, Pritchard RH, Jiménez-Sánchez A. (1991) A calcium-binding protein that may be required for the initiation of chromosome replication in *Escherichia coli. Res Microbiol* **142**: 137–140.

Zaritsky A, Woldringh CL, Pritchard RH, Fishov I. (2000) Surviving *Escherichia coli* in Good Shape: The Many Faces of Bacillary Bacteria. In, J Seckbach (ed.), *Journey to Diverse Microbial Worlds*, Kluwer Acad Publ: The Netherlands, pp. 347–364.

One of Bob's last pictures while talking about science — over a pizza, Amsterdam, September, 1999. All 4 coauthors of his last article, with Dutch colleagues.

Left to right: Conrad Woldringh, Robert Pritchard, Anton van Geel, Peter Huls; Nienke Buddelmeijer, Nanne Nanninga, Arieh Zaritsky and Itzhak Fishov.

Itzhak Fishov is gratefully acknowledged for preserving this precious photograph.

An Overview

Arieh Zaritsky* and Peter A. Meacock†

Bob Pritchard was influential in advancing several fundamental concepts of Genetics and Microbial Physiology during his early career in the 1950s in Glasgow, under the mentorship of **Prof. Guido Pontecorvo**, and subsequently in Hammersmith, London, as a postdoctoral researcher with **Prof. William (Bill) Hayes**. His early achievements included the demonstration that genes are linear in structure and that mutational changes can occur at many sites, so challenging the previously-considered "atomic" indivisible nature of the Mendelian gene; these conclusions, and others about the placement of cross-over events between chromosome pairs, were reached from his detailed investigations into the process of recombination in the filamentous fungus *Aspergillus nidulans*. It was probably while in Hammersmith that Bob's curiosity about the control of DNA replication and the coupling of chromosome duplication with cell division in bacteria was triggered. His fascination with this topic was further stimulated during his collaborative work with **Prof. Karl G. Lark** in Kansas USA, and the investigation of these fundamental processes occupied the rest of his active academic life.

This compendium begins with a tribute and summary of Bob's contributions in science and local politics, by **Peter A. Meacock**, a PhD student

*Ben-Gurion University of the Negev, Israel (ariehzar@gmail.com) and Peter A Meacock, University of Leicester, UK (mea@le.ac.uk).
†Department of Genetics, University of Leicester, LEI TRH (mea@le.ac.uk)

(1971–75) with Bob. It is followed by a brief description of the "embryogenesis" of the Genetics department at Leicester University by **Sir Hans. Kornberg FRS** and two accounts of the first years of the department by **Susan E Wilkins** and **Peter T. Barth**; both were the first PhD students of Bob, and amongst the small number of those recruited in 1964 to the newly established Leicester Genetics department by one of the then youngest life sciences professors in the UK. These are followed by portrayals from **Drs Barry Holland**, later promoted to become a second Professor at the department, and **William (Bill) Grant**, then a research fellow (1970–74), who transferred from Genetics to join the newly established Microbiology department in the brand new Leicester Medical School, where he was promoted to a Professorship.

Aspects of the years preceding Bob's appointment to establish the now-famous Genetics department are briefly described by his American peer **Karl Gordon Lark**, in whose laboratory at Kansas State University crucial experiments were jointly performed, dissociating the initiation and elongation components of the complex chromosome replication process. **Abraham Eisenstark**, also an academic visitor at KSU, had daily association with Bob **during** this time and then spent a Sabbatical year in his newly established Leicester department. **Marilyn Monk** was contemporary of Bob in Hayes' MRC Microbial Genetics Unit at Hammersmith Hospital, London, also spent a short period at Leicester. In contrast, **Millicent (Millie) Masters** and **William (Willie) Donachie** were also members of the Hammersmith group and significant contributors in the same research areas as Bob, but followed independent parallel pathways eventually securing positions in Edinburgh, a situation that led to friendly rivalry and productive collaboration. The accounts from these peers of Bob in their youth make a fascinating read, while being instructive to current and future generation(s) of scientists.

Six of Bob's other research associates have written about their experiences with Bob during the first two decades of the Leicester Genetics department: **Shamim I. Ahmad** (1966–72), **Ifor R Beacham** (1967–71), **Arieh Zaritsky** (1969–71), **Michael (Mick) Chandler** (1970–73), **Ramon Diaz Orejas** (1975–78) and **Alfonso Jiménez-Sánchez** (1982–84). It is noteworthy that four of these came from overseas and received excellent treatment in line with Bob's liberal outlook, and contradictory to misplaced perceptions about conservatism of English society. The summaries of their experiences illustrate Bob's attributes: an incisive analytical mind, the capacity to integrate fields, a holistic view of "the cell", close participation in performing actual experiments when other commitments allowed, and a willingness to consider non-conventional ideas.

The young department (see photograph above) was endowed with a relaxed atmosphere that resulted in long-term friendships, joint grants and articles, both then and in later years. **Douglas Smith** describes the sabbatical period that Bob spent in San Diego just before officially retiring. The careers of both **Mick Pocklington** and **Vic Norris** testify to the lasting influence of Bob after his retirement; Mick as a postdoctoral researcher with **Eli Orr**, another of Bob's Postdoctoral Associates and Senior Lecturer who passed away recently, and Vic as an EMBO Fellow with Barry Holland who later was promoted to a Chair at Rouen University, France. The last, but by no means least, of Bob colleagues to add his memoirs and appreciation is **Sir Alec J. Jeffreys FRS**, who was recruited to the Department of Genetics by Bob. The scientific achievements of Sir Alec have contributed greatly to the outstanding international reputation of the department.

A chapter by **Peter Meacock** and **Annette Cashmore** summarises the history of the Genetics Department after Bob retired, and identifies some of the research highlights and other individuals who have played key roles in its successful development. This is followed by a contribution by from **Arnie Gibbons** and **Iain Sharpe**, two of Bob's Liberal Democrat colleagues who record his contributions to local politics at both City and County levels.

Throughout much of the time he was establishing the Department of Genetics at Leicester, and developing his research reputation, Bob was a single parent raising two sons (**John** and **Simon**), later joined by **Naomi** whom he adopted when he married his second wife, **Susan**. The loss of both sons before his own passing was a terrible tragedy for Bob, but having **Naomi** and her young family with him during that difficult period must have given him comfort and consolation. Tragedy again struck Bob for the last decade of his life, when he suffered a neurological condition that left him incapacitated and unable to communicate. **Naomi Matthews-Pritchard** has kindly contributed the final chapter of this volume, in which she describes Bob's personal and family life.

Most of Bob's scientific life concerned advancing our understanding of the fundamental aspects of bacterial cell duplication (see list of his major publications above), merging physiological, genetic, biochemical, computational and microscopical approaches as necessary for any particular question asked.

EMBO Workshop: "Incompatibilty and Control of Replication in Plasmids", Sønderborg, Denmark, 1978.

Partial List of participants, projected linearly from *left-to-right*): Petter Gustafsson, Walter Messer, Bob Pritchard, Kurt Nordström, Soto Hiraga, Werner Maas, Jenny Broome-Smith, Hilary Richards, Dietmar Blohm, Millie Masters, Willie & son (David) Donachie, Michael Yarmolinsky, John Nijkamp, June Scott, Bob Rownd, Fleming Hansen, Bruce Kline, Joachim Frey, Knud Rasmussen, Lucien Caro, Dean Taylor, Dick Novick, Jim Wechsler, Marilyn Nugent, Kaspar von Meyenburg, Jorge Crosa, Peter Meacock, Ken Timmis, Kenichi Matsubara.

Photographed by Peter T. Barth (and see back cover page for a B/W variation).

Contents

1 Bob Pritchard: Eminent Scientist and Remarkable Individual

Peter A. Meacock*

Robert Pritchard was an eminent Geneticist, founder of the department where DNA-fingerprinting was discovered, and a prominent local Leicester Liberal Democrat politician. He was held in high esteem by his scientific peers worldwide and respected by both his political colleagues and adversaries. To his friends and colleagues he was always "Bob".

He was born on January 25th, 1930, the youngest of three children to Florence (Stella) and Henry Pritchard OBE, an Assistant Commissioner of the Forestry Commission. His childhood was spent in Wandsworth, south-west London, with evacuation during the war years to Radstock, Somerset. His high intelligence was quickly recognised and after grammar school in Somerset he attended the Emanuel School in Battersea where he gained several O-levels with distinction and passed the Higher Schools Certificate within one year, instead of the usual two. He was accepted to Kings College, London, to read Botany and was awarded a First Class degree and the Carter Prize for Physical Biology; during these formative years his passion for biology and genetics was kindled.

His PhD research was carried out in Glasgow under the supervision of Professor Guido Pontecorvo FRS, on genetic recombination in the filamentous fungus *Aspergillus nidulans*. He was one of the first people to map

* Department of Genetics, University of Leicester, LEI TRH (mea@le.ac.uk).

the order of mutations within a gene, directly paralleling the more widely cited work of Benzer on the T4 bacteriophage r_{II} genes, and so proving the concepts of linearity of the gene and that mutations could occur at any position. He also studied the positions of the cross-over events responsible for genetic exchanges between chromosomes and was the first to document "Negative Interference" between recombination events; these were ground breaking discoveries in the 1950s, just after Crick and Watson had solved the structure of DNA.

After his PhD he joined the MRC Microbial Genetics Unit led by Professor William (Bill) Hayes FRS at Hammersmith Hospital, London, and shifted his interests to the newly emergent field of bacterial genetics and continued his recombination work using *Bacillus subtilis* transformation by naked DNA as his system of choice. He subsequently spent time at Kansas State University in the USA with Professor Karl G. Lark where he started the research for which he is best known, namely the control of chromosomal DNA duplication and its coordination with cell division; he also studied plasmids, mini-chromosomes in bacteria that are often the agents conferring antibiotic resistance. In 1964, at the remarkably young age of 34, Pritchard was recruited to the University of Leicester as Professor, with the remit to establish a Department of Genetics as part of a new School of Biological Sciences. Instrumental in that was Professor (now Sir) Hans Kornberg FRS, Head of Biochemistry, who had been so greatly impressed by Pritchard while taking the 1960 Hammersmith course on Microbial Genetics that he persuaded the then Vice-Chancellor of Leicester, Charles (later Sir Charles) Wilson, to establish a federal School of Biological Sciences in order to provide an integrated biological sciences degree course, and appoint Pritchard as the Head of Genetics.

Throughout the time he was building the department and serving as Head, Pritchard continued to run a large research group which focussed on understanding the mechanisms that regulate the commencement of DNA duplication (initiation) in bacterial cells and how that process is co-ordinated with cell growth and division; topics of fundamental biological importance and central to understanding what happens when our own cells grow uncontrollably in cancer. Amongst the many scientific papers that were published, an early one (1969) was identified to be one of the most widely cited by other researchers and particularly influential as it challenged existing dogma in the field. That proposed a model of negative control on DNA initiation, "The Inhibitor Dilution Model", wherein the number of chromosomal replication origins where DNA replication is initiated is in some way titrated against the cell mass, and the process of initiation can only take place when certain conditions are fulfilled. The model postulated is a biological clock that determines the periodicity of DNA replication during cell growth and division.

Other important studies by his group addressed the biosynthesis of the thymidine nucleotide, a building block unique to DNA, and showed how limitation of the supply of this important metabolite could modulate the overall rate of DNA synthesis and change the relative dosages of genes within the cell in ways entirely consistent with the new idea that the circular bacterial chromosome is replicated in a bidirectional manner from a single origin site. Subsequently, Pritchard's group went on to investigate the properties of mutations in genes encoding components of the chromosome initiation machinery and to devise methods for the recovery of molecular DNA clones of chromosome origin regions.

Because of his clear thinking and willingness to discuss ideas with others without prejudice or bias, Pritchard was admired and respected by the international scientific community from postgraduates to Nobel Laureates. He was widely sought after as a research collaborator both in the UK, particularly by the cell cycle researchers in Edinburgh and London, and in other countries, notably in Denmark, Holland, Israel, Spain, Sweden, Switzerland and the USA. He was invited to speak at many scientific meetings, national and international, and his contributions were unanimously acknowledged as being both original and provocative for furthering the field of research endeavour. His arguments were always logical and rational, and based on critical thinking about the experimental data.

Pritchard was an inspiring teacher and mentor. His lectures to undergraduates in all years of their degrees stimulated students, and challenged them to understand the experimental evidence supporting the topics they were being taught. As a research supervisor his enthusiasm and passion for his science was infectious; he would spend long periods in the laboratory with his PhD students and postdoctoral associates discussing their findings in great detail and encouraging them to think carefully about the conclusions from their data and to explore new ways of addressing fundamental questions. When a new experiment was underway he would often be alongside his student to see the results coming off the instrument, keen to be involved in their analysis.

His academic contributions were not limited to his own research and teaching. He served as Chairman of the School of Biological Sciences (1977 to 1979) and sat on various committees of the University and grant awarding bodies where he gave wise counsel. However, he was frustrated when he saw mindless and blinkered bureaucracy by government and other science policy-makers. He spoke out strongly, writing letters to national newspapers and international periodicals and even appearing on the TV programme *Controversy* in 1974, about issues where he felt incorrect decisions had been made without full consideration and discussion of the evidence — particularly the moratorium on recombinant DNA experimentation of the 1980s when he felt the conjectured hazards of "genetically modified organisms", and the proposed actions, were totally contradictory to logical conclusions based on hard scientific evidence.

He was concerned too, about the increasing prevalence of antibiotic resistance in bacterial infections, brought about by the indiscriminate over-use of those drugs in human and veterinary medicine during the 1970s and since. He argued from fundamental Darwinian principles that the carriage of resistance imposes a burden to the cell that only confers advantage in the presence of the particular antibiotic as the selection agent. By temporarily withdrawing use of individual antibiotics under a tightly controlled antibiotic-use policy, resistance to the drug would eventually disappear from the bacterial population. However, his arguments fell on deaf ears and we now have the life-threatening problems we face today.

Other topics he wrote about included strategies for the prevention of the spread of AIDS, the involvement of patients in the choice (even administration) of their own care pathways, and the moral and ethical decisions relating to surrogacy and *in vitro* fertilisation. He advocated early and open discussion of controversial societal issues, arguing that authoritarian regimes could often misinterpret facts and subsequently impose inappropriate policies which in turn contributed to public mistrust of science and its policy-makers. Instead, he believed passionately in public participation by informing individuals of the facts related to an issue, so empowering them to make their own judgements and voice their opinions.

The Department of Genetics that he founded with its modest complement of only seven, flourished under his leadership, with his reputation as a scientist attracting new staff and collaborators from around the world. He established his department on true democratic, community and collegiate principals. Academics were encouraged to focus on what they did best — curiosity-driven research and teaching; he allowed them free rein in selecting the topics and directions of their research and rarely intervened. All members were treated equally, and everyone from the Head down to technical staff and students in laboratories was on first name terms. Large-scale research instruments and facilities, and even the consumables budget, were operated communally and available to all; this meant that staff who did not currently hold a research grant were able to maintain progress on their project until further funding was obtained.

Pritchard quickly convinced the University to provide additional academic positions for the department and the initial group was expanded by appointment of staff with interests addressing a range of topics in genetics and molecular biology: fungal, developmental and population genetics (respectively Clive Roberts, Jennifer Dee and Robert Semeonoff); antibiotic proteins and cell membranes (Barry Holland); plasmid biology (Brian Wilkins). This coincided with the Department's move (1967) into new accommodation in the Adrian building where two features were introduced that were central to its vitality and success: a communal tea-room, the site of much heated scientific (and political) discussion, generally with Pritchard at its centre, and a centralised media kitchen (for glassware wash-up, growth media preparation and sterilisation) staffed by dedicated technicians to support both research and teaching programmes.

The establishment of the Medical School at Leicester in 1975 was a significant development at the University in which Pritchard again showed his astute judgement, and prevented his department from stagnating in an increasingly difficult financial climate. He was influential in the design of the medical teaching curriculum and ensured that genetics was included in the pre-clinical training of all medical students, which brought opportunity to expand the staff complement. This allowed him to introduce further dimensions to the department's portfolio, including microbial pathogenicity (Peter Williams) and mammalian genetics (Grahame Bulfield).

Pritchard had immediately recognised that the new recombinant DNA technology, pioneered in the USA by Berg, Cohen & Boyer and others, offered enormous possibilities for fundamental and applied genetics research in both biology and medicine and he sought to recruit staff experienced in these techniques. Amongst those recruited (1977) was a young Dr. Alec, now Professor Sir, Jeffreys, and so the molecular era of genetics research was launched at Leicester eventually leading to the major discovery of DNA fingerprinting. Pritchard saw DNA fingerprinting not only as great science from his own department, but also as an important judicial development that could help protect the rights of the individual.

Today the department that he founded has expanded still further and is recognised as one of the foremost Genetics Departments in the UK, producing research of international significance and continuing to operate with the same ethos and philosophy.

In 1983 Pritchard relinquished the headship of the Department of Genetics in order to spend more time on his other abiding passion, politics. He was a life-long member of the Liberal, later Liberal Democrat, Party and was elected firstly in 1987 to Leicester City Council as a Liberal Alliance candidate, and later to Leicestershire County Council as a Liberal Democrat. He represented the East Knighton division, became leader of the Liberal Democrat Group on Leicestershire County Council, and was a continual challenger of the policies of the dominant Labour group. Despite their disagreements he was respected by the other parties because of his courtesy, intelligence and willingness to listen to other opinions. He cared about improvement of the urban environment, advocating policies that could lead to resettlement within city centres, and campaigned vigorously on a variety of local issues including the preservation of the wonderful Victorian facade of the Leicester railway station and trees within the city.

He was a humanist who embraced mankind irrespective of race, religion or cultural background. He fought injustice with the belief that equal rights were the most logical result of any consideration of the human condition. While on academic study leave in the 1970s at the University of California in San Diego, California, he frequented a beach bar and befriended some of the local "drop-outs". Pritchard became disgusted by the immorality of the American police force and was appalled when one of the bar's "regulars" was accused by the police of criminal behaviour purely on prejudice, with no real supporting evidence. However, those officers had not anticipated the presence of a British Professor who was brave enough to act as witness. Pritchard spoke eloquently against the prosecution and the defendant was acquitted.

Many years before the current "fat cat" controversy Pritchard was a holder of shares in a major UK manufacturing company that had grossly under-performed and so was proposing to freeze the pay of, or even lay off, a considerable proportion of its workforce. To the astonishment of many, it was simultaneously proposing to award the Chairman a substantial bonus for "good management". Pritchard spoke publically at the AGM against the motion, asking how these two proposals could be reconciled, arguing that the company was not treating all of its employees equally and fairly and that one of its functions was to create employment; his objections were reported and discussed by the national press.

Bob Pritchard greatly enjoyed the company of other people and relished discussions, sometimes deliberately playing "Devil's Advocate" in order to test the validity of his own arguments and to explore all possible avenues that he and others might have overlooked. He was a generous man, supportive and caring to those close to him and a hospitable host, throwing large parties for the department at Christmas/New Year. When he moved to his new house, which had a swimming pool in the garden, he opened it up to his academic and political colleagues during the summer months — a gesture that was delightedly taken up by postgraduates and younger staff!

In 2002 he was struck down by a neurological condition that left him almost totally incapacitated and unable to communicate, requiring long-term hospitalisation until his death on 12th April 2015. It was a tragic and cruel end for somebody who had contributed so much.

Bob Pritchard truly was a remarkable man with great clarity of thought and the courage to speak out against prevailing opinions and dogmas when he felt they were wrong.

Peter A. Meacock was a PhD student with Professor Pritchard, 1971 to 1975, and is now Emeritus Reader, Department of Genetics, University of Leicester.

This article is a modification of an Obituary of R. H. P., written by P. A. M. and originally published on the University of Leicester web site (http://www2.le.ac.uk/staff/community/people/tributes/obituary-for-professor-robert-hugh-bob-pritchard). It was also used as the basis for articles about RHP in the "Leicester Mercury" on April 16, 2016 (http://www.leicestermercury.co.uk/Tributes-paid-eminent-Leicester-scientist-Liberal/story-26343494-detail/story.html) and "Times Higher Education" on May 7, 2015 (https://www.timeshighereducation.com/news/people/obituaries/robert-pritchard-1930-2015/2020006.article).

2 How It All Began: The Genesis of Genetics at Leicester

Sir Hans L. Kornberg FRS*

1960 was, for me, an *annus mirabilis*. Quite unexpectedly (unexpectedly by me, that is), I had been offered the newly created Professorship of Biochemistry at the University of Leicester. This university had, only relatively recently, graduated from its previous status as a University College, founded shortly after WW I as a war memorial and affiliated to London University. Its former Principal, now Vice-Chancellor, Dr. (and future Sir) Charles Wilson, had become acutely aware that, although the University had, over many years, offered courses by the small Departments of Botany and of Zoology, oriented largely towards ecology and taxonomy, there was no adequate teaching in Biochemistry or Molecular Biology, and instruction in Genetics rested solely on the shoulders of a distinguished Reader but overworked fungal geneticist, Dr. John Fincham. In my interview with the Senate committee seeking to remedy these deficiencies, and in my Inaugural Lecture, I had outlined grandiose plans for the University to establish a School of Biological Sciences — plans which, to my gratified astonishment, were enthusiastically supported by most (if not all) members of the Senate. Since all life processes depend on the selective expression of genetic potential, and as John Fincham was about to leave for the John Innes Institute, such a School needed as a priority to establish an independent Department of Genetics.

*Boston University, 5 Cummington Mall, Boston, MA 02215 (hlk@bu.edu).

By a happy coincidence, I was at precisely this time a student in a Summer School in Microbial Genetics, run by that wonderfully gifted teacher Bill Hayes, Head of a MRC Unit at the Hammersmith Hospital, where a young (and equally gifted) teacher, Bob Pritchard taught us ham-handed but eager pupils the rudiments of yeast genetics. It became blindingly obvious to me that Bob would be just the person needed to launch the new Department on its path of teaching and research. I put this idea to Bob; he came to visit me and proved receptive to the proposed developments; Charles Wilson set the bureaucratic machinery in motion and, in 1964, Bob was duly installed as Professor of Genetics and Head of the Department.

I had the great good fortune to work with Bob for the subsequent 11 years, and to participate with him in the creation of an effective School of Biological Sciences and the design of the new Adrian Building. But it is not only as a former colleague that I remember him. Bob became and remained a dear friend to me and to my family and it is as that we shall always treasure his memory. *In piam memoriam*, Bob.

Raising a glass. Professor Guido Pellegrino Arrigo Pontecorvo (R) and Robert H Prichard (Center) with friends and colleagues during a Glasgow Genetics department outing to Loch Lomond, 29 June 1964. (*University of Glasgow Archive Services, Guido Pontecorvo Collection; GB248 UGC 198/10/1/1/17.*)

3 Early Years of the Genetics Department, University of Leicester

Susan E. Wilkins (Hollom)*

Together with Margaret Peake, Shamim Ahmad and Shaleen, I was in the first group of staff recruited by Bob Pritchard for the new Genetics Department in 1964. Bob interviewed me at Hammersmith Hospital, where he was based before coming to Leicester. Bob himself was appointed by Hans Kornberg, then head of Biochemistry. The new department superseded the geneticists in the Botany Department, rather to their surprise. Academic staff were Bob Hedges and Barry Holland. Clive Roberts and Jenny Dee came later. So the nascent Department of Genetics at the University of Leicester, including the kitchen/preparation woman, had 9 members. From acorns mighty oaks grow *etc.*

We were housed in the Victorian R-Block, behind engineering, while the Adrian building was built. To get to Bob's office you had to go past Margaret, who typed on a state of the art electric typewriter using tippex correction fluid when necessary and occasionally ripping pages out of the carriage with a satisfying tearing sound. Margaret thought she would just be a clerk when she arrived on the first morning and discovered that she was now the professor's private secretary.

There was an open lab where the research students worked, a washing up area and autoclave and side offices for the lecturers. I was appointed

---

* Oadby, Leicester (susan.e.wilkins@btinternet.com).

as Technician-in-charge. Bob took a leap of faith in giving me this job as I knew nothing about organising a lab, having previously worked on vaccine production in other people's research labs. Later, Bob suggested I work towards a PhD, which I got in 1969. Bob patiently read my attempts at thesis writing and helped me to improve.

Bob insisted on giving lectures to the first year students to start them off in what he considered to be the correct way of thinking about genetics.

There were many visitors to the department including Bill Hayes from Edinburgh, John Subak-Sharpe from Glasgow and, of course, Hans Kornberg. Bob would often start an argument for the sake of it. To me, he seemed an independent thinker, a leader rather than a follower. I remember Bob saying to an aunt of mine that he did research to prove his brain was working properly!

We had a good social life. On hot days we would decamp to the local open-air swimming pool, now closed. We had department parties in the lab with decorations from Kirby and West (which had to be returned) and beer brewed on the spot and "cleared" through the continuous flow centrifuge just before being drunk, still rather cloudy. Bob gave many parties in his own house, which was on Knighton Road initially. Later, after Bob had moved to Knighton Grange Road, parties centred round the swimming pool.

The move from R block to the Adrian building was overseen by Clive in the Adrian with me still in R block. Clive pulled rank and I was unable to prevent the coffee machine from leaving before I did.

Once in our new home, Terry Lymn, a properly trained technician, was appointed.

We had many visitors from Israel, America and Sri Lanka. Research was on microbial and fungal genetics at that time.

Then Brian Wilkins arrived from Yale. Brian was soon to become my husband and later to be head of department. Brian and I used to visit Bob at home frequently where the discussion would centre on politics and the stock market. Telex with the latest stock market movements was nearly always showing on the TV. Latterly, before his tragic illness, Bob became a city councillor for the Liberal party and tried valiantly to stop the development at Hamilton, north of the city. In this he failed.

Some of "Bob's People" (1964–1978) at the 50th Anniversary of the department, In front of the Adrian Biology Building, October 11th, 2014.

Sitting (from *left*): John Collins, Arieh Zaritsky, Peter T. Barth (with his photos-album), Susan Wilkins (Hollom), Margaret Cullingford (Peake), Ramón Díaz Orejas.

Standing (from *left*): Grantley Lycett, Maria-Elena Fernandez Tresguerres, Peter A Meacock, Michael (Mick) Chandler.

(Photograph courtesy of Department of Genetics, University of Leicester.)

Alec Jeffreys put the Department in the public eye with his discovery of Genetic Finger printing.

I hope Bob is aware that the Department has grown exponentially since its foundation in 1964 and now is rated the best Department of Genetics in the country.

4 Bob Pritchard and the Control of DNA Replication in Bacteria

Peter T. Barth*

After over 50 years of research, the mechanism of control of DNA replication in *E. coli* is still not fully understood. Professor Robert (Bob) Pritchard made a major contribution to the thinking and research on the problem. He was the founding Chair of the Department of Genetics in Leicester University in 1964. I joined him as a PhD student in that year, although this came about by chance. I had been accepted for a PhD place by Professor Sir Hans Kornberg in the adjacent Department of Biochemistry. During the summer vacation I received a letter from Hans telling me that the newly appointed Professor of Genetics was looking for a PhD student. Hans already had many students so he asked if I might be interested in working on DNA replication with Bob Pritchard instead of on a biochemical project. My BSc Hons degree had been in Physiology and Biochemistry at Southampton University with very little course work on bacteria and molecular biology. However, I had read about the structure of DNA and early work on genetics in *E. coli* which I felt was a hot topic of future research. So, much as I admired Hans Kornberg, I opted for a place with Bob, which I have never regretted.

Bob played a leading role in studying the problem of how the replication of bacterial chromosomes is controlled. He spent some time in

*Global Cancer and Infection Research Department, AstraZeneca, Alderley Park, Cheshire SK10 4TG, UK (retired) (peter.barth@talktalk.net).

Karen Ippen-Ihler, Bob Pritchard and Kaspar von Meyenburg in discussion at the Plasmid Workshop Meeting in Berlin, April 1978.

Ole Maalϕe's lab at Copenhagen in the 1950s. For many years they measured cell size and the chemical composition of *Salmonella typhimurium* cultures under various growth conditions. They also studied the effects of changes in growth conditions,[11,12,19,20] Bob also did a postdoctoral year in Karl Lark's laboratory in Kansas State University just before accepting the Chair of Genetics at Leicester. There they showed that permitting cell growth (of *E. coli*) while preventing DNA replication by thymine starvation of a thymine auxotroph, stimulated replication when thymine was restored.[15]

Bob took me under his wing and showed me the basics of bacterial genetics and DNA analysis by ultracentrifugation. Robert Hedges, lecturer in the department was also very helpful in teaching me molecular biology. He had an amazing recall of the literature and was always willing to stop and explain things to me. I continued the above project (from Lark's lab) using thymine starvation and other methods of stopping DNA replication, including nalidixic acid, the function of which was not known at the time. The results were consistent with the hypothesis that the capacity of a cell to reinitiate replication at the origin rises exponentially in step with the increase in mass. In my last year in Leicester, I visited Dr. Willie Donachie at the MRC lab, Royal Postgraduate Medical School in London. He told me he had deduced that the cell mass per origin at the time of initiation was constant but didn't tell me how he had done it. Using the relationship established by Schaechter *et al.*[19] between cell mass and growth rate, the cell age distribution[14] and the time of initiation of chromosome replication in relation to the cell mass[6] I was able to calculate for myself that the mass of a cell per origin at the time of initiation (m_i) was a constant. (John Collins and I only recently discovered that Bob later asked John to do this calculation without telling him that I had already done it. It was available in my thesis, but Bob may have mistrusted my maths.) It is salutary that this conclusion, which played a major part in the development of models of the control of replication, turns out to be only an approximation. It was shown later,[22] by measurement of the cell mass per origin at the time of initiation using flow cytometry, that over a 5-fold decrease of growth rate there was a 1.5-fold increase in m_i.

At this time, most workers in this field, including Willie Donachie,[7] supported the notion that replication control was positive, i.e. an initiation factor accumulated during cell growth and when the concentration reached some critical level, it triggered a round of replication. Bob, however, developed a negative control model. He proposed that an inhibitor of initiation (H) expressed by a gene close to the origin, blocks initiation until its concentration has halved by cell mass increase. At this time, initiation is triggered, causing the gene to express a "pulse" of H which blocks re-initiation for a full cell cycle until [H] has halved again. The model has interesting self-regulatory properties which fit with the observed narrow range of cell sizes found in Maalφe's lab.[20] The publication of this work[17] became a "Citation Classic". It contradicted the model of Jacob *et al.*,[10] which postulated a specific attachment site for each replicon harboured by the host bacterium. This was postulated to explain incompatibility between such replicons. Bob pointed out that the phenomenon of incompatibility can be explained without specific attachment sites as a logical consequence of replicons sharing the same replication or segregation mechanism. The Jacob model also became less credible when it was realised that a bacterial host could harbour very many plasmids together. In later years, we constructed a strain of *E. coli* containing eight compatible plasmids and showed that each maintained the same copy number as it did when it inhabited the strain alone.[4]

In 1965 Jane and I got married. We had a large flat with two spare rooms let out to other students. I let it be known that if anyone had any furniture they wanted to be rid of, we could do with some. Bob gave us a very useful leather armchair that went in our sitting room. He was very welcoming to Jane who was far from "home" in Brighton. At one of our occasional parties, he showed us his party trick of opening a bottle of wine without a cork-screw (no screw-tops in those days). He took the foil off the neck, wrapped a tea towel around the base of the bottle and hit it against the wall several times. Each hit forced the cork out a little by the inertial impact of the wine, until he could take hold of the cork and pull it out. Warning — don't let visiting professors do this in your home, it may lead to broken bottles! Eighteen months later Jane gave birth to our first child, Nicholas. This was the first baby conceived in the department, so to speak. Bob was tolerant of my absences and was very welcoming to our baby. In my last year Bob informed me that a postdoctoral position at Yale with his friend and colleague Prof. Paul Howard-Flanders was available. Of course, I applied. I met Paul in London when he was over for a visit. Bob also suggested that I should apply for a Fulbright Scholarship. (For this I wrote a research proposal to discover the function of nalidixic acid. Sadly, I made no progress. Several years later, Nick Cozzarelli discovered that it inhibited DNA gyrase.) So Bob not only helped me enormously to start in this area of science but was also very helpful in furthering my career.

The use of thymine-requiring mutants in the DNA replication experiments led Bob to instigate research on the nature of the mutations concerned. Stimulated and guided by Bob, this was a fruitful collaboration between Ifor Beacham, Shamim Ahmad and myself.[1,2,5] In *E. coli* the first mutation knocking out thymidylate synthetase (*thyA*) led to a requirement for high levels of thymine in the growth medium but secondary mutants arose easily which required much lower levels. We investigated the nature of these mutations. They were found to have a defective catabolism of deoxyribonucleotides leading to the excretion of deoxyribose. They fell into two categories which we showed were mutations in the deoxyriboaldolase or deoxyribomutase genes. This blocking of the catabolism of deoxyribose-1-phosphate led to the intracellular conversion of thymine to thymidine which could then be further metabolised and utilised in DNA synthesis.

I remember Bob telling me that the trouble with conferences is that everybody reaches more or less the same conclusions and they then go back home and all do the same experiments. However, he did organise conferences himself. I particularly remember the one he arranged with Kurt Nordström on Incompatibility and Control of Replication in Plasmids in Sønderborg, Denmark in 1978. Bob also told me that it is worth thinking hard and doing fewer, more significant experiments rather than ploughing on repeating what you are already doing. On the other hand, he had a tendency, I think, to underestimate the difficulty of solving certain problems, replication control included. For a short time in Yale I was working on the breakdown of DNA in *recA* strains. Bob wrote in one of his letters to me (Sept 1969), "I'll buy you a bottle of whisky if you can solve what RecA does." I think a whole case if not a small Scottish distillery would have been more appropriate.

Bob was not slow to criticise scientific dogmas when he thought they were wrong. I was involved in plasmid research in Naomi Datta's lab when the Asilomar Conference on Recombinant DNA was held in February 1975 in California. Paul Berg was a leading proposer of the conference. He had cloned monkey virus (SV40) into phage lambda and was worried that it would get into *E. coli* and then infect the human population. As a result of the conference, stringent regulations came into force in the US around cloning. Not to be outdone, the UK had its conference on "Recombinant DNA" in Oxford in July 1975. I was invited to give a paper in the first morning session: "Transposable Elements & Recombination Plasmids" (sic) as we were the first to discover the phenomenon of the transposition of drug resistance genes in plasmids.[3,9] Having explained transposition, I then discussed how, in principle, plasmids can pick up DNA from any source and distribute it to other bacteria and that, whatever genes give them a survival advantage will be retained by Darwinian selection. Considering the vast human and bacterial populations, is it likely that our little efforts in the lab could produce dangerous bacteria that could not evolve in the outside world? But before I could finish my point, I was cut off by the chairman as "we were running out of time" but I think it was more that I was running outside the favoured dogma. In the evening discussion session we were asked to vote for or against the proposed regulations to "safeguard the public against the dangers of cloning". Bob Pritchard stood up and pointed out that these "dangers" were a matter of speculation. And since when has any scientific question ever been solved by voting? Is the world round or flat? It would have been voted "flat" by most people for most of history. Even in the USA today, how would people vote on the question, "How did the living world arise: via evolution by natural selection, or by divine creation"? Bob said that what was needed was research to test these supposed dangers, not votes based on ignorance or prejudice. He later published a paper in *Nature*, "Recombinant DNA is safe".[18]

His colleague in Leicester, Hans Kornberg, obviously held similar views on this issue. There was a meeting in Cardiff in 1988 on "The Release of Genetically-engineered Micro-Organisms".[21] I gave a talk which I dedicated to Bob Hedges who, tragically, had died earlier that year. In his opening remarks Hans said, "Whenever I consider the ability of Man to alter to any significant and longstanding extent the interplay of the myriad genes already present in the environment, I remember Aesop's fable of the fly that sat on the axle of the wheel of the chariot and said, 'Look at the dust I raise'. Only time and the results of the type of experiments that we are already carrying out, and that will be the substance of our discussions here, will tell."

Bob also criticised the overuse of antibiotics which has led to a rapid and seemingly inexorable rise in the prevalence of drug-resistant bacteria. He proposed a cyclical use of antibiotics[16] such that when one hospital found rising resistance to drug A, it switched to drug B until resistance to drug A had subsided. This was actually rather naive as bacteria do not live in little enclaves and the selection pressure of over-use of drugs affects populations at large. Also, different drugs are indicated in different infection situations. If drug A was on the "banned" list but this was the only agent to which a pathogen was sensitive, does the physician use A or risk the life of his patient? The Swann Report published in 1969 recommended that antibiotics used in human medicine should not be used in agriculture, especially the widespread use of sub-therapeutic levels of antibiotics as growth promoters. It was only in 2006 that this practice was banned in the EU. It continues in most of the rest of the world and involves the use of tons of these precious agents which produces a continuous selection pressure for the spread of drug resistances. Bob was also aware that plasmids (including F) were capable of picking up genes from the chromosome and from cohabiting plasmids, leading to the development of multi-resistant plasmids which could then spread rapidly between bacterial species.[16]

In the summer of 1980, I went to San Diego for a three month "sabbatical" in Don Helinski's laboratory (UCSD, La Jolla campus). Jane and our two children (Caroline and Nicky) came with me. I was working on the mechanisms of stable maintenance of the promiscuous plasmid RP4. This involves multimer resolution, segregation as well as replication control. As noted in Doug Smith's and Naomi Pritchard's chapters, by a happy coincidence, Bob, Naomi and Suzi were also there at that time. We met up several times, in Doug's lab with Judith Zyskind and others for scientific discussions, on the beach and at various houses for social occasions. Beach parties were very common and successful because in California you could always rely on good weather. We gave a party at our condominium one evening and Bob turned up very late and very giggly. He explained he had been smoking pot and had lost track of time. Bob's attitude to drugs was that they should all be legalised but controlled by education and taxation. Alcohol and nicotine were legal through historical accident, not because they were harmless. He thought marijuana was less harmful than these two drugs and that it was an injustice to penalise users. Politicians are afraid of the subject and I doubt whether he made any headway with his Liberal colleagues. Attitudes are changing slowly, however. Suzi was very involved with theatre there. We went to see a very memorable production of Shakespeare's *Love's Labour's Lost* given in the open air. This photo was taken in Balboa Park, which is huge and full of wonderful things — a concert platform, museums, the Botanical Building, theatres, lakes and gardens.

The negative regulator H proposed in the Pritchard *et al.* model[17] does not exist in *E. coli* or in any other bacteria, as far as we know. However, all plasmid replicons examined to date have negative feedback control mechanisms, which is consistent with the model. Kurt Nordström, who was a close colleague of Bob Pritchard, worked on the plasmid R1. He showed that replication is initiated by a Rep protein. The expression of the mRNA for this protein is controlled by an antisense RNA. As this RNA is unstable, its concentration is always proportional to the copy number of the plasmid and hence gives negative feedback control.[13] The other more common mechanism for plasmid copy number control is based on iterons: multiple repeats of a particular DNA sequence. The mechanism for this is complex, involving binding of the Rep protein and steric hindrance of the *ori* region but basically, it is also a negative feedback mechanism.[13]

Caroline, Jane, Bob, Nicky (behind Bob) and Suzi in Balboa Park, 1980.

The mechanism of control of DNA replication in *E. coli* has turned to be far more complicated than Bob or anyone else envisaged in 1969. The replication origin *oriC* and the initiator protein DnaA are still the main players but multiple mechanisms have been shown to control the level of the active initiator DnaA-ATP by auto-repression, titration at binding sites, by conversion to the inactive DnaA-ADP and by sequestration by hemi-methylated DNA. Most people working in this field have concluded that the control is positive, in the sense that when sufficient DnaA-ATP has accumulated, initiation will occur. A recent paper from Kirsten Skarstad's lab[8] shows that a modest increase (50%) in the cellular concentration of DnaA-ATP does not change replication control. This proves that although DnaA is the initiator,

Bob Pritchard in his Nursing Home on 11 October 2014, the day of the 50th Anniversary celebration of the founding of the University of Leicester's Genetics Department.

it is not the limiting factor. Control may still prove to be by an unrecognized feedback inhibitor. I think that would have surprised and pleased Bob.

References

1. Barth PT, Beacham IR, Ahmad SI, Pritchard RH. (1968) Properties of bacterial mutants defective in the catabolism of deoxynucleosides. *Biochem J* **106**: 36–37.
2. Barth PT, Beacham IR, Ahmad SI, Pritchard RH. (1968) The inducer of the deoxynucleoside phosphorylases and deoxyriboaldolase in *Escherichia coli*. *Biochim Biophys Acta* **161**: 554–557.
3. Barth PT, Datta N, Hedges RW, Grinter NJ. (1976) Transposition of a DNA sequence encoding trimethoprim and streptomycin resistances from R483 to other replicons. *J Bacteriol* **125**: 800–810.
4. Barth PT, Richards H, Datta N. (1978) Copy numbers of coexisting plasmids in *E. coli* K12. *J Bacteriol* **135**: 760–765.
5. Beacham IR, Barth PT, Pritchard RH. (1968) Constitutivity of thymidine phosphorylase in deoxyriboaldolase negative strains: Dependence on thymidine requirement and concentration. *Biochim Biophys Acta* **166**: 589–592.
6. Cooper S, Helmstetter CE. (1968) Chromosome replication and the division cycle of *E. coli* B/r. *J Mol Biol* **31**: 519–540.

7. Donachie WD. (1968) Relationship between cell size and time of initiation of DNA replication. *Nature* **219**: 1077–1079.
8. Flåtten I, Fossum-Raunehaug S, Taipale R, Martinsen S, Skarstad K. (2015) The DnaA protein is not the limiting factor for initiation of replication in *Escherichia coli. PLoS Genet* **11**: e1005276.
9. Hedges RW, Jacob AE. (1974) Transposition of ampicillin resistance from RP4 to other replicons. *Mol Gen Genet* **132**: 31–40.
10. Jacob F, Brenner S, Cuzin F. (1963) On the regulation of DNA replication in bacteria. *Cold Spring Harbor Symp Quant Biol* **28**: 329–348.
11. Kjeldgaard NO, Maalφe O, Schaechter M. (1958) The transition between different physiological states during balanced growth of *S. typhimurium. J Gen Microbiol* **19**: 607–616.
12. Kjeldgaard NO. (1961) The kinetics of RNA and protein formation in *S. typhimurium* during transition between different states of balanced growth. *Biochim Biophys Acta* **49**: 64–76.
13. Nordström K. (1990) Control of plasmid replication — How do DNA iterons set the replication frequency? *Cell* **63**: 1121–1124.
14. Powell OE. (1956) Growth rate and generation time of bacteria with special reference to continuous culture. *J Gen Microbiol* **15**: 492–511.
15. Pritchard RH, Lark KG. (1964) Induction of replication by thymine starvation at the chromosome origin in *E. coli. J Mol Biol* **9**: 288–307.
16. Pritchard RH. (1966) Contagious Drug Resistance in Bacteria. *J Roy College of Surgeons of Edinburgh* **11:** 123–125.
17. Pritchard RH, Barth PT, Collins J. (1969) Control of DNA synthesis in bacteria. *Microbial Growth: Symp Soc Gen Microbiol* **19**: 263–297.
18. Pritchard RH. (1978) Recombinant DNA is safe. *Nature* **273**: 696.
19. Schaechter M, Maalφe O, Kjeldgaard NO. (1958) Dependency on medium and temperature of cell size and chemical composition during balanced growth of *S. typhimurium. J Gen Microbiol* **19**: 592–606.
20. Schaechter M, Williamson JP, Hood JR, Koch AL. (1962) Growth, cell and nuclear divisions in some bacteria. *J Gen Microbiol* **29**: 421–434.
21. Sussman M, Collins CH, Skinner FA, Stewart-Tull DE. (1988) The Release of Genetically-engineered Micro-organisms. Academic Press. ISBN 0-12-677521-4.
22. Wold S, Skarstad K, Steen HB, Stokke T, Boye E. (1994) The initiation mass for DNA replication in *Escherichia coli* is dependent on growth rate. *EMBO J* **13:** 2097–2102.

5 Exciting Times in Leicester with Bob (Building the Commune)

Barry Holland*

Imagine the scene: in 1964, Leicester University finally has a Department of Genetics — founded by young man, 34 years old, a hot shot in the new world of bacterial molecular genetics. Bob Pritchard, a dashing and highly intelligent fellow had arrived. Unfortunately the University could only spare three or four smallish rooms in a tiny corner of a rather old building, with the penitentiary-like name, the "R-block". Seven souls, including Bob and two PhD students, were the early inhabitants in year one (see Susan Wilkins chapter) and apparently with not much of an endowment. The staff swelled to 10 in 1965 with my arrival plus a student and a technician. After that it was just one exciting expansion after another: expensive new buildings, a burgeoning teaching staff, undergraduate and graduate students, postdocs, technical and administrative personal. Thus evidently, Bob successfully mobilized support for this new discipline within the University, establishing an exciting department that had already achieved a strong international reputation by the mid-70s.

Imagine too that in this era, although we had the exquisite structure of DNA and some glimpses of the synthesis of DNA, we had little or no idea of how replication (or the whole cell cycle) was regulated. However, Bob himself, together with Karl Lark (1964, JMB) had just provided the first

*University of Paris-Sud 11, France (ian-barry.holland@u-psud.fr).

evidence that replication of the chromosome of *E. coli* stemmed from a unique origin. At the same time we were all still digesting the revolutionary concept of the operon, promoters and repressors (Jacob and Monod 1963), and the accompanying rise of molecular biology (first issue of JMB, 1959). Moreover, the gold rush of results in 1961 to 1963 had only just established the information flow from DNA to RNA to Proteins and the nature of the genetic code. Nevertheless, the detailed mechanism of replication and even the first replication initiation mutants, and how do plasmids and bacteriophages replicate, still lay in the future. Similarly, nucleic acid hybridization, cloning, recombinant DNA technology, the third kingdom of life, membrane protein crystal structures, PCR, biotechnology, nucleic acid sequencing methods, the secrets of membrane structure and function, the extraordinary mechanism of bacterial cell division and much more were still to be discovered. What a time to teach and research in biology and Bob Pritchard gave me that opportunity!

I met Bob for the first time towards the end of summer 1963 in my lab in DD Woods Microbiology Department in Oxford. I was in my third year with an independent postdoctoral fellowship (i.e. I supervised myself, a bit like my PhD). In retrospect it seems that Bob out of nowhere just dropped in on me one day, talked about the exciting prospects in his new Department in Leicester, and chatted over what I had been doing in Oxford, what were my future plans — very informal, very friendly. I had only 3–4 publications at that point all concerning *Bacillus megaterium* bacteriocins, considered to be extracellular inhibitors of essential processes in competing bacteria. I was able to say that I had discovered a new megacin that inhibited cell division and induced degradation of DNA. However, I was determined to switch my research to *E. coli* and acquire new skills in molecular biology, as I was shortly to take up a postdoc position in Champaign-Urbana with Sol Spiegelman. Bob was very interested in my plans and said he would keep in contact with the possibility of my joining him later in Leicester. I was a stunned to hear that but if that was how people were recruited, why not! Bob, happily for me, was true to his word and near a year later he wrote to offer me a junior lectureship. Things were going extremely well in Urbana, with our successful test-tube replication of Qβ-RNA, but I discussed the offer with Sol who knew Karl Lark and had heard great things about Bob Pritchard and Sol encouraged me to accept.

A few months later I arrived in Leicester in September 1965, but no formal interview with a University committee of worthies — I just started. In addition, Bob miraculously had organized for me a first class graduate student (Caroline Hill) and an excellent technician (Kristina Bellamy). I discussed with Bob what my role would be and in the style that would always be his hallmark — in essence — No pressure, No interference, Do the research that you wish to do, teach primarily second and third year students the unfolding fundamentals of replication and protein synthesis. Ask the encyclopaedic Bob Hedges for his notes to get started! He also hoped, although I do not remember Bob's precise comments, that I would take on the responsibility to introduce the molecular biology techniques into the Department that I had seen and picked up in Urbana. So I was given a free reign and it worked! For the next 20 years that Bob continued, with barely a break, as Head of the Department, my teaching and especially my research evolved or changed radically several times until I had sampled many of the major areas of the emerging fundamental biology of *E. coli*. Throughout that time Bob provided his full support when I needed it.

As I have said on many occasions, the spirit in the Department from the outset was of a liberal democracy, the communal budget not least, quite a contrast to laboratories in the States, the UK and Europe that I have known. This gave us the freedom to take risks, the freedom to do what we wanted, both in teaching and research. We had a thoroughly democratic process for assigning Departmental grants for PhD training. The training program itself was way ahead of its time, with students giving one departmental talk per year, and preparation of a detailed first year report. The community spirit also expressed itself with always extremely well attended seminars, many departmental social events. These included the wonderful summer time frolics in Bob's so generously provided, open house swimming pool not far from the campus, plus football (indoor) and cricket teams, even the raising of a healthy tomato crop in the south side of the new Adrian building. All this was a great formula for the success and expansion that followed at every level. Not to be underestimated was the focal point of the working day where everyone who wished could meet anyone and everyone in the coffee/tea room established in the Adrian Building — invariably packed, standing room only, at least for the mid-morning session. In the general hubbub, new people were introduced, advice on anything could be sought, problems were solved, ideas exchanged, the politics of the day of course and even sport discussed (although the latter was not a favourite topic of Bob's).

Bob and I had many coffee break discussions (with blackboard illustrations), they could be heated but never rancorous, we debated the role of the membrane (or its non-role in Bob's view) in division and DNA replication, even harder discussions over whether the DNA regulator was likely negative or positive. We also had strongly differing views on the moratorium on recombinant DNA. Of course we 'fought' hardest over politics but neither of us gave an inch over the differing policies of the Liberals and the Labour Party, although he could be very persuasive. In those discussions I learned a lot about Bob's great analytical mind, his rigorous way of doing research (which I could never quite emulate) and his contribution in particular to the emerging understanding of DNA replication and its control. People of the stature of Francois Jacob and Jim Watson admired his work and were very happy to visit him in Leicester with all the rest of us benefiting. Obviously, Karl Lark and the Scandinavian and Dutch cell cycle Schools, including of course Kurt Nordström, were all great admirers. Some other memories include Bob being invited to appear in a BBC TV panel (around 1974) concerning controversies at the interface of science and society. A number of us went down to London to be in the audience and I remember Bob attacking the poor (or non-existent) regulations supposedly to avoid misuse of antibiotics and presenting in his logical and powerful way what should be done about it. Thinking about all of this induced me to re-read Bob's chapter in the book "Genetics and Society" (published in 1993 by Addison-Wesley, Publishers Ltd.) covering the events of the De Montfort Hall Symposium for the 25-Year Anniversary of the Department in 1990. The Chapter, entitled "Public participation in the scientific adventure", is a gem, a brilliant analysis of where the "authorities" go wrong in dealing with antibiotic use but also, on the poor National policies on AIDS advice, having a child via surrogacy, how really to do personalized medicine (patients controlling own pain relief dosages), why not to be afraid of GMOs, and genetic and ethical issues generally. It is well worth reading (should be available still from Pearson Education, www.pearsoned-ema.

com; tel +44 (0) 1279 623623). This sums up Bob Pritchard very well, what he thought, how he analyzed a problem and how he selected a likely solution although not the one that most people would have thought of. In this chapter there is also more than a hint of Bob's impatience, dissatisfaction with Governments, and with terrible irony, Doctors.

Life within the Bob Pritchard Department was, it is important to say, never dull — the exciting science, the community spirit, exemplified perfectly by the coffee room culture. We had lots of important and interesting visitors, bringing exciting seminars and discussions. As a consequence we were well connected with the outside world, and for those of us involved in that, a near perfect teaching load. Moreover, within a couple of years of establishment we were planning the details of the laboratories in the new Adrian building. Then only a few years later we were in expansion mode again with the building of the Medical School. Bob worked very hard, supported by all the staff, to convince the medical people that a single Department of Genetics should contribute the teaching for medical students, providing them with the strongest fundamental basic science, and in the literal sense a bridge between science and medicine. As far as I remember the Dean of Medicine, Robert Kilpatrick, crucially was a key convert to this concept.

Finally, of course Bob, aided by the staff, was not alone in these adventures. This was a time of radical change in academic science in general in the UK and in Leicester in particular. But Bob immediately saw the opportunities and embraced the changes whole-heartedly. The culture that he believed in and promoted, encouraged hard work but also risk taking — stemming from the freedom to think without pressure from above, a culture that invited us to think "let's do something new today". The same culture also produced a superbly organized Unit, with key positions of laboratory Manager (novel in those days) and Departmental Secretary, respectively Terry Lymn and Margaret Peake, both hand picked by Bob. This ensured, for over 30 years at least, the smooth running of the whole operation, even in some extremely difficult financial times.

Bob Pritchard, an outstanding individual with a great intellect, strong and dynamic personality, made his mark both as a creative scientist and as a political force in the wider community. In science he will be long remembered for his ground-breaking studies on the regulation of DNA replication. But let us not forget his immensely important contributions to the structure of the gene and the mechanism of recombination from his work in Glasgow in *Aspergillus nidulans*, right at the beginning of his career. I shall remember him too for the freedom he gave me, and many others, to learn without fear and the opportunity to express ourselves to the best of our abilities.

Cutting the Silver Jubilee Cake

Photo courtesy of Leicester Mercury

From Left: Professor Robert Pritchard, Professor Barry Holland, Professor Alec Jeffreys, Dr Peter Williams

6 Memories of the Genetics Department at Leicester

William (Bill) Grant*

I had been on my first postdoc as a Fulbright travel scholar in Madison Wisconsin, at the McArdle Laboratory for Cancer Research where I was working on the slime mould *Physarum polycephalum*, which was then considered to be an ideal cancer model system in which to study the control of mitosis. Coming back to the UK was facilitated by the SRC "repatriation fellowships" that were relatively easy to get because at that time the UK government was worried about the large number of scientists leaving for the USA never to return. Deciding where to go in the UK was a no-brainer since not only did Leicester have a *Physarum* slime mould group with Jenny Dee in Genetics, there was also a *Dictyostelium* slime mould group with John Ashworth in Biochemistry. When I arrived in Genetics in May 1970, the academic staff comprised Bob plus Barry Holland and Brian Wilkins all working on aspects of *E. coli*, Jenny's slime mould *Physarum* group, Clive Robert's *Aspergillus* group and Robert Semeonoff working on population genetics of field voles. Robert, coincidentally, was my second cousin, although we had never met before, having followed parallel paths at different schools and different university courses in Edinburgh.

I had never heard of Bob until I arrived in Leicester but he immediately made me feel welcome, giving me my own laboratory and office

*Department of Infection, Immunity and Inflammation, University of Leicester (wdg1@leicester.ac.uk).

(just imagine that happening today!) and the department was extremely supportive to my efforts to set up a research programme. I quickly came to admire Bob's remarkable intellect — he had an incredible capacity to strip problems down to basic, often simple questions. While in Wisconsin I had learned how to grow dinner plate-sized *Physarum* vegetative plasmodia where all the millions of nuclei in the syncytium went through mitosis at exactly the same time and I brought this technology to Leicester to complement Jenny's work on the amoeboid sexual stage. Clive's graduate student Pete Fantes and Jenny's student Alan Wheals were already interested in the cell cycle in eukaryotes and I began to supervise a new PhD student, Pete Sudbery, using the *Physarum* system to look at what might trigger mitosis. I had been deeply impressed listening to Bob describe very simple but clever experiments with *E. coli*, where results (of any kind) invariably provided useful insights. Underpinning these were the questions posed by thoughtful models of what might be going on. We clearly needed a model (or models) for what might be going on in our system and obviously Bob was the man to talk to about models. As a consequence of a few weeks discussion, usually in the tea room, we developed a series of experimental approaches to be used by Pete Sudbery, based on some preliminary thoughts about possible models for the control of mitosis. As a consequence of these discussions, we eventually later published a theoretical paper on cell cycle control in eukaryotes which is still cited today and is amongst my most cited publications (PA Fantes, WD Grant, RH Pritchard, *et al.* (1975). The regulation of cell size and the control of mitosis. *J Theoret Biol* **50**: 213–244).

There was a great sense of camaraderie in the department in those days with much out of hours socialising. The department then shared the first floor of the Adrian Building with Hans Kornberg's Biochemistry department. A great rivalry developed between the two departments over the years, including cricket matches, an athletics contest and even a chess tournament where the deciding match was between Bob and Hans, which they cleverly contrived to draw!

It is worth recording how supportive Bob was to his staff in the early department and how they often went on to great things (a testament to Bob's ability to spot talent). There is also the famous story of the one that got away — in about 1974 Genetics advertised a lectureship in eukaryote

Sunday Lunch in The Cradock Arms pub, Commemorating Bob, October 12th, 2014.

From left: Michael (Mick) Chandler, Peter A. Meacock, Chris Dodd, Grantley Lycett, Maria-Elena, Fernandez Tresguerres, Jean Emeny, Susan Wilkins (Hollom), Margaret Cullingford (Peake), Ramón Díaz Orejas, Pilar Ferraz-Pascau, Peter Fantes, Susan Grant (Armitt), William (Bill) Grant, Arieh Zaritsky. (Photograph courtesy of P. A. Meacock.)

genetics. Peter Fantes was then a postdoc together with Paul Nurse in the lab of Murdoch Mitchison (the doyen of the cell-cycle people at that time) and they both applied. Murdoch said he couldn't distinguish between them so they both got shortlisted. Obviously I knew Pete well and I also knew Paul so they stayed with us (for some reason we had only one operational guest bedroom at that time so we set up a bed in the dining room and they tossed up for the bedroom — Paul lost!) Neither of them got the job — yes, the department had such a high collective opinion of itself that it turned down the future Sir Paul Nurse FRS Nobel Laureate! It was some consolation I suppose that the guy who got the job, Graham Bulfield, went on to head up the institute that cloned Dolly the sheep!

I eventually got a temporary lectureship in 1973, still physically in Genetics but with the understanding I would move to the newly set up Department of Microbiology based on Peter Sneath's MRC Microbial Systematics Unit just as soon as the new Medical School building opened on the old Adrian Building car park — this I duly did in late 1974 and spent the rest of my career there in various reincarnations of the original Microbiology department, eventually getting a chair. On the way, I completely changed research field and went back to my roots in microbiology because we couldn't make cell cycle mutants like Leland Hartwell (Paul Nurse's co-Nobel Laureate) was doing in yeast, because *Physarum* was seriously polyploid even in the sexual stage! Over the years doing all kinds of microbiology, I always remembered Bob's advice to sit down and think about the reasons for doing experiments before rushing into them — what does a positive result really mean, plus just as important, what does any negative result equally well tell you? Wise council!

7 Skating on Not-So-Thin Ice in Kansas

Karl G. Lark*

Bob Pritchard and I were contemporaries in every sense of the word. Both of us were born in 1930 and we both had entered the budding field of molecular genetics, albeit by different routes. Our paths intersected in a most unlikely place, the middle of Kansas where we collaborated on experiments that helped to define the origin of DNA replication as a special region of the *E. coli* chromosome. This is the story of that collaboration.

It had its origin in the early 1950's when I was finishing my thesis with Mark Adams at NYU (Lark and Adams 1953) and Bill Hayes, a hitherto unknown microbial geneticist was becoming known for his pioneering work on Hfr matings (Hayes 1952). On the way to my postdoctoral in Copenhagen we stopped and visited with Bill in London where he was beginning to attract young people into a group that was to become the "Hammersmith" group. In 1961, my wife Cynthia and I spent a sabbatical in Scotland and, while there, again visited Hayes' new group in London. I am not sure when I first met Bob, but it was probably then. Certainly, by that time I had become aware of Bob Pritchard as a serious force in genetics.

By that time I had begun to focus my research on cell division and the initiation of DNA replication. Ole Maalϕe and Phil Hanawalt had pub-

*Distinguished Professor Emeritus, Department of Biology, University of Utah, Salt Lake City, Utah 84112, USA (Lark@bioscience.biology.utah.edu).

lished data on the amount of DNA synthesized by bacteria in which protein synthesis had been blocked (Maalφe and Hanawalt 1961). They suggested that this corresponded to finishing chromosome replication before starting a new cycle of replication; and that although protein synthesis was needed for initiating a replication cycle it was not needed for completing an ongoing replication cycle. We decided to test this hypothesis using a double labeling technique (Lark *et al.* 1963).

Using a bacterial auxotroph requiring multiple amino acids as well as thymine (*E. coli* $15T^-$), we starved the bacteria for essential amino acids for a period corresponding to more than a generation time, until DNA replication had ceased. Upon restoring amino acids we labeled the first DNA synthesized with tritiated (H^3) thymine. After more than a generation of growth, we again starved the bacteria for essential amino acids and restored amino acids substituting 5 bromo-uracil for thymine, thus synthesizing density labeled DNA. Analyzing the DNA thus formed in a CsCl density gradient we demonstrated that the H^3 DNA was preferentially synthesized with a density label — i.e. replication in the absence of protein synthesis repeatedly ceased at a particular region of the chromosome, presumably the origin of replication.

This experiment was originally tried when I passed through Maalφe's laboratory toward the end of my sabbatical in Scotland in 1961. I explained the experiment that I proposed to Ole and asked if I could carry it out while passing through his lab. It seemed appropriate, given his initial insight. So I placed the density labeled bacteria in tubes to be centrifuged in a swinging bucket rotor in order to equilibrate in a CsCl gradient, and I departed on a short visit to Stockholm returning a few days later to remove the tubes and analyze their content. The centrifuge appeared to have been stopped and its placidity infuriated me. I angrily demanded to know who had stopped my "run" and where had they put my tubes. I was assured that no one knew anything about what had happened. Upon opening the "Spinco" I was greeted by disaster: the rotor had exploded and demolished the drive shaft and the interior cooling system, but the debris was contained by the armored cylinder that enclosed the centrifuge chamber. Using my available grant funds I was able to arrange for the repair of Ole's centrifuge and leave his lab with some degree of amicability. When I returned to Scotland I recounted the experience to Avrion Mitchison, a friend in the

King's buildings. He replied "I suppose you could blow up any number of centrifuges on those terms." We did just that during the period when we successfully carried out such experiments later, in our laboratory in St. Louis.

My collaboration with Bob Pritchard began around 1963 shortly after I moved from St. Louis University to Kansas State University. The techniques we had used to blow up Maalφe's centrifuge and "demonstrate an origin" of replication (density plus radioactive labeling in a pulse chase protocol), could also be used to test an idea of Bob's. Shortly before my move to Kansas, at a meeting in Aspen Colorado, Bob had hypothesised that following a period of thymine starvation to block DNA synthesis, addition of thymine led to resumption of synthesis at the existing replication fork *but replication also was initiated by a new fork at the chromosome origin — two replication forks on the same template*. In those days, it was an accepted practice to collaborate when testing someone else's idea rather than seeking to "beat them to it". In keeping with this custom, I invited Bob to come for a year and collaborate on testing his idea.

So Bob joined us in Manhattan Kansas. If my memory serves me, his marriage was breaking up and he brought his two young sons with him (or they joined him shortly after he came) as well as a sister who created a home for the four of them in the middle of the Kansas wheat country. He lived in an apartment separated from our laboratory by a set of tennis courts and it was a simple matter for him to walk home and check on the children. I think that this aspect of life in Manhattan, together with the fact that it was a small town, provided a relief from the stress of previous years. Certainly he enjoyed the time with his children. Another aspect that appealed to Bob was that our laboratory was situated in a building that contained the departments of physics and mathematics, a social and physical ambiance that he enjoyed.

Thus it was at Kansas State University that we did the experiments that proved his idea to be correct (Pritchard and Lark 1964):

Using *E. coli* $15T^-$, we starved cells for essential amino acids for more than a generation period. We then reinitiated growth in H^3 thymine for a short period after which we removed the radioactive label and grew the cells for several generations in non-radioactive medium. We then starved the culture for required thymine while allowing protein synthesis. After more than a generation of thymine starvation we added 5-bromo-uracil to reinitiate replication. The preferential replication of the H^3 DNA (incorporating density label) demonstrated the initiation of replication at the same site as the one reached during replication in the absence of protein synthesis. The fact that replication was initiated from the same region of the chromosome after two such quite different metabolic treatments provided the strongest evidence available at that time to support the hypothesis that this region contained the origin of replication. Later, similar experiments using a shift up into "rich" growth medium demonstrated that yet another treatment also initiated new replication forks from this same region of the chromosome (Bird and Lark 1968).

Outside of the laboratory Bob was active and one winter day came in to the laboratory and told us that he had been skating on Tuttle Creek, a lake close by (15 minutes' drive). This changed the lives of all of us. Winter became not only bearable, but fun and those of us who had never been skating (myself included) learned to skate and joined a group of assorted people that had fun on ice three feet thick where we skated, watched over by eagles sitting in dead trees immersed in the lake, or at night by moonlight to the accompaniment of the cries of coyotes. Although the science was a major step forward, I will always remember Bob most gratefully for making Kansas' winters bearable by introducing us to skating "in the wild", admittedly an unlikely aspect of the Bob Pritchard that most people knew.

References

Bird R, Lark KG. (1968) Initiation and termination of DNA replication after amino acid starvation of *E. coli* 15T$^-$. *Cold Spring Harbor Symp Quant Biol* **33**: 799–808.

Hayes W. (1952) Recombinations in *E coli* K-12: unidirectional transfer of genetic material. *Nature* **169**: 118–119.

Lark KG, Adams MH. (1953) Stability of phages as a function of the ionic environment. *Cold Spring Harbor Symp Quant Biol* **18**: 173–183.

Lark KG, Repko T, Hoffman EJ. (1963) The effect of amino acid deprivation on subsequent DNA replication. *Biochim Biophys Acta* **76**: 9–24.

Maalφe O, Hanawalt PC. (1961) Thymine deficiency and the normal DNA replication cycle I. *J Mol Biol* **3**: 144–155.

Pritchard RH, Lark KG. (1964) Induction of replication by thymine starvation at the chromosome origin in *Escherichia coli*. *J Mol Biol* **9**: 288–307.

8 Reflections — Bob Pritchard

Abraham Eisenstark*

Our friendship started when Bob [and sister and two sons] came to Kansas State U. Manhattan, Kansas, when Bob joined Gordon Lark's lab for some key experiments. I was a faculty member and Bob, Gordon and I had considerable scientific interaction, but also many social conversations. However, I was already aware of Bob's contributions to the flow of genetic concepts that were emerging in that era.

The origin of Bob's and Gordon's research interest was perhaps the interesting investigators and their results that were emerging from Ole Maalϕe's Laboratory. Earlier, I had been on a Guggenheim Fellowship in 1959 in Maalϕe's Lab, a witness to the discoveries of genetic regulation of bacterial growth concepts that were under investigation.

After Bob's activities in Manhattan, our friendship continued the following year in Leicester, where I was on my second sabbatical. I participated on a research project.

I also developed a keen friendship with Shamim Ahmad, in Bob's lab in the old building. Shamim and I have maintained continual collaboration in both England and USA until this day, collaborating on thymineless death studies, etc. In Leicester, I also enjoyed deep friendships with Bob Hedges [what a sharp mind!!].

---

*Byler Distinguished Professor [emeritus] University of Missouri, Columbia. Mo (EisenstarkA@missouri.edu).

Imagination of something makes it real — Pablo Picasso

Some Leicester Genetics "Elders", 43 years later, at the 50th Anniversary of the department, in front of the Adrian Biology Building, October 11th, 2014.

Standing (from *left*): Peter Fantes, Roger Buxton, Barry Holland, Michael Chandler, Tony Samson.

Sitting (from *left*): Arieh Zaritsky, Peter T. Barth, Susan Wilkins (Hollom), Margaret Cullingford (Peake), Jenny Foxon, John Collins.

(Photograph courtesy of Department of Genetics, University of Leicester.)

But all my interactions with all of the members of Bob's office and lab staff were such a joy. We went out for Indian lunches, swimming on the first warm day in summer, evening pubs, etc, the Wiggeston Boys School next door to the lab [our two sons attended, our daughter went to Collegiate Girls School].

This was also an opportunity to present seminars and to visit distinguished molecular microbiologists in other parts of GB, even the high-security prison next to Hammersmith. [I accidentally walked through a gate next to Hammersmith, which turned out to be the prison.]

And with my family, on weekends, we gorged ourselves with visits to area historic sights and to live theatre plays in nearby cities [Nottingham, etc.].

As for my personal activities, my broad interest in bacterial genetics lured me into several niches, including phage studies, and mutations in *S. typhimurium*, but also biannual UNESCO committee trips to London's Royal Academy. Our meeting room had huge portraits of great British scientists. To my great surprise, I kept staring at that great British scientist, Benjamin Franklin.

The following two decades were filled with much microbial genetics activity: five full summers on Long Island USA with family while working in Miloslav Demerec lab, yearly trips to international meetings mainly in Europe but also Australia, Israel and Japan. I also was at U. Paris for a few months with Daniele Touti working on bacterial superoxide dismutase genes. I also got involved in administrative jobs as National Science Adm. Program Director for Molecular Biology, and a decade as Director of Division of Biological Sciences at U. Missouri.

Years later, I had mandatory retirement from U. Missouri in 1990 [at age 71] but was invited to be the Director, Cancer Research Center, Columbia, Mo. Thus, for the past 25 years, including today, I have been in the lab working on *Building a better Mousetrap: using microbes to target and destroy tumors.* Our research utilizes some concepts developed in Bob's lab.

Finally, I must now conclude to shift to reading the proofs of a chapter in the volume, edited by Shamim Ahmad.

I am truly pleased to read this set of memoirs. It has already led to sharing experiences with others who were once associated with Bob. Nice to recall memories of associations with Barry Holland, Susan Hollom and John Collins.

9 Remembering Bob at the Hammersmith and Leicester in the 60's

Marilyn Monk*

Bob Pritchard was a terrific guy. I have often thought of him over the years although I had lost touch as I had moved away from *E.coli* in 1972 into slime moulds and, then, in 1975, into early mammalian development. I was sad to hear that Bob had become so ill and that he had suffered such loss in his family.

I first met Bob in 1961 when I arrived at Bill Hayes' MRC Microbial Genetics Unit situated, as part of the London Postgraduate Medical School, at the Hammersmith Hospital. I was sent to Bill Hayes to study for my PhD by Bruce Holloway, my BSc Honours and MSc supervisor in Melbourne. I had already spent several years in Melbourne (1959–1961) in the field of microbial genetics, isolating and characterizing the bacteriophage and pyocins of *Pseudomonas aeruginosa* and studying lysogeny and transduction, and the influence of irradiation. From these studies, I was lucky enough to be awarded an Australian scholarship and free passage (five weeks on a ship) for study overseas.

Arriving in London to Bill Hayes' Unit in 1961 was an awe-inspiring experience. There I met senior scientists including — in addition to Bill Hayes — Bob Pritchard, Roy Clowes (my immediate supervisor), Ken Fisher, Stuart Glover, Jeff Shell, Ken Stacey and Neville Symonds. They

* Institute of Child Health, London (m.monk@ucl.ac.uk).

seemed to me a whole generation older and wiser than myself, although the age difference was only about 10 years. Bill Hayes also organised occasional meetings with the Cambridge group — Francis Crick, Sydney Brenner, John Smith and others — which I found quite scary. My fellow PhD students at the time were John Scaife, Julian Gross, Donald Ritchie, Faith Poole, Michael Kelly and Paul Broda. I was very much alone arriving in England and missing my family and friends and my horse. I lived in a strictly-supervised college for academic ladies in Chelsea and I was a ward of the High Commissioner of Australia who organised lots of treats for me — Lord Mayor's Balls, visits to Buckingham Palace garden parties and a box at the Royal Albert Hall. This was of course all very new and strange for a girl from the Aussie bush and it also appeared rather strange, I think, to my fellow students. However, Bob would not have been fazed by my incongruous life style. Everybody in the lab was kind and supportive and helped me to forget about the Australian sunshine, my horse, barbeques and the surf, and teach me how to be more worldly-wise and cultured. Bob always seemed to me to be the clever senior person in the unit and the most open-minded in his approach to science and the world in general.

My studies in London followed on from my work in *Pseudomonas* — mainly with transmissible colicin factors. I returned with my PhD to Melbourne in 1964, but after a year working on recombination in phage T4, I was restless and wanted to return to Europe. I won another scholarship — the ICI_ANZ scholarship for study overseas — which was considered by some to be most unfair — especially being a female! This time I plane-hopped back to France to post doc at CNRS Gif-sur-Yvette with Raymond Devoret, continuing with my interest in transmissible elements and induction of bacteriophage lambda. In Paris, we worked on the cross-induction of lambda (Borek-Ryan effect) with irradiated *colI* factor, thus indicating that the mechanism of activation of the bacteriophage lytic cycle was initiated by a piece of irradiation-damaged DNA.

It was from Paris that I went to work with Bob in Leicester for a year's post doc (1966–1967), continuing with observations on the mechanisms of indirect induction under his supervision. I was aware of Bob's passionate interest in politics and it was a political time for me as well with my activities with fellow Australians against the Vietnam war. My work during my brief time in Leicester was not so much concerned with Bob's interest in control of DNA replication and cell division. However, after leaving Bob's lab, his influence became evident in my next position back at the Hammersmith Hospital in London with Bill Hayes.

Howie Goldfine, then visiting Bill Hayes in London, had isolated 500 temperature-sensitive mutants of *E. coli*. He made these available for me to isolate 45 temperature-sensitive mutants in DNA replication simply by analysing the differential uptake of ^{3}H thymine (measuring DNA replication) and ^{14}C leucine (measuring protein synthesis) after shift to high temperature. Although Bob had departed to Leicester, many of the original researchers from the early 60's were still at the Hammersmith and now Millie and Willie Donachie, and Jim Shapiro had joined us. Then, in 1968, most of us moved with Bill Hayes to Edinburgh when his MRC unit was amalgamated with that of Martin Pollock in the new Molecular Biology Department at Kings Buildings in Edinburgh. I continued research with some the DNA mutants I had isolated in London together with further work on DNA repair and the newly defined Kornberg polymerase repair enzyme. However, after a few years, I had become rather bored with bacteria not doing much of interest that one could watch. So I started working on chemotaxis in *E. coli* and, after hearing an inspirational talk by James Bonner, I changed field to slime mould signaling and aggregation. With the closure of Bill Hayes' unit at his retirement in 1975, I was rescued by Anne McLaren and I had to change field again. This time I made the enormous leap — requiring the development of single cell molecular biology — to molecular studies of mouse and human embryos in Anne's MRC Mammalian Development Unit in the Galton lab in London. So, for the last 40 years I have been outside of the field of DNA replication and repair, and so I had lost touch with Bob.

I liked and admired Bob enormously and I remember clearly the main influences he had on my ongoing research. I remember him telling me NOT to read the literature too much as current dogma would stop me from discovering anything new. That must have influenced me because I did a lot of discovering of strange things that no-one else believed — such as the 'anti-Weismann' late origin of the germ line in mammals (1983), the deprogramming methylation erasure in mammalian preimplantation development and primordial germ cells (1987), and single cell molecular biology.

I learnt from Bob to "think outside the box". I also remember Bob having an influence in teaching me that thinking was more important than doing. Many a long night I have spent not sleeping and thinking about my research. Then the answer seems to come from without — often serendipitously and by inexplicable "mistakes". I will always be grateful to Bob and glad to have known him.

10 The Real Geneticist, Already at Bill Hayes' MRC Unit

Simon Silver*

I knew Bob Pritchard well but only briefly, arguing science and the world daily, during my 2 years in Bill Hayes' Medical Research Council Microbial Genetics Unit, which I visited in 1960 — I now see Bob was already there — then at Hammersmith Hospital, London, 1962–1964, and am pleased to add my memories of the best of times and places, where he expressed himself strongly about good science and the world around us.

The Unit was extraordinary in many ways, some of which Bob carried with him to Leicester, as best described in Peter Meacock's deep Obituary, posted at www2.le.ac.uk and repeated here (pp. 12–17). Since I am now more than 50 years out of date, I have very little useful to add.

Everyone at the Unit was on a first name basis, including Bill, Bob, Neville Symonds, Ken Stacey, Julian Gross, Marilyn Monk, Don Ritchie and Paul Broda, all of whom went on to hold Chairs in British universities. Roy Clowes, Ken Fisher (earlier) and I came to the US. I certainly left out some names and some such as John Scaife continued wonderful microbial genetics studies without the large burden of administrative work that many took. Bill's MRC unit incubated Professors like the Cambridge MRC Molecular Biology unit incubated Nobel Prizes.

* Microbiology and Immunology, University of Illinois at Chicago (simon@uic.edu).

Bob was a real geneticist in his thinking, while a few like me were more molecular biologists with genetics only as the core thinking. Some of us were more focused on experimental science, while Bob included a strong theoretical genetics interest, which helped in the broad subject matter he later encouraged in Leicester. Some of us stayed solidly in microbial genetics research, while others changed primary for secondary interests, as Bob Pritchard did.

Our politics ran a spectrum, but my views were quite close to Bob's. My memory is that he had strong interests in local government already then.

11 Me and My Beloved Professor, Dr. R. H. Pritchard

Shamim I. Ahmad*

I came to Bob's (Professor R.H. Pritchard) laboratory in 1966 as a research technician with the possibility of registration for a PhD at the University of Leicester. Then, the Department of Genetics was fairly new and as I learnt I was the third PhD student after Drs. Peter Barth and Roger Young (Dr. Young left before my arrival). Then came Dr. Ifor Beacham and we were sharing the same laboratory and were more or less working as a team, yet independently on various metabolic and catabolic aspects of DNA synthesis. In early days I worked (as a trainee) under the supervision of a Post-doctoral fellow, Dr. Marilyn Monk, and 3 months later Bob took me under his supervision and assigned me a program of work to select mutants of *Escherichia coli* by using selective pressure by 5-fluorouracil + deoxyadenosine. It did not take long to isolate an expected mutant and its enzyme analysis showed that it was missing purine nucleoside phosphorylase. In another joint study with Drs. Barth and Beacham, we discovered that dRib-5-P was the inducer of deoxynucleoside phosphorylases and deoxyribose aldolase. Two papers were jointly published in 1968, one in *Biochim Biophys Acta* and one in *Biochem J* which gave me (us) the psychological boost resulting in my subsequent publications of another 4 research papers (see the publication list of Pritchard, R.H.) before I submitted my

* Nottingham Trent University (shamim.ahmad@ntu.ac.uk).

thesis for the PhD degree. Subsequently, I left the Genetic Department and, as a lecturer, joined Trent Polytechnic which later became Nottingham Trent University.

Now in 2016, while looking back at my early life in UK as well as my studies at Bob's laboratory I feel I was exceptionally lucky to meet Bob at the time when I was desperately looking for an academic position. I approached Bob on the phone and after an interview he offered me the position of a research technician under his supervision which I was "over the moon" to accept alongside a reasonable salary from his MRC grant.

Additional luck favored me that the subject (bacterial physiology and genetics), which I was working on in those days was considered fairly valuable. This is evident from the fact that as soon as I completed my education at Leicester, I applied to different Universities and research centers (McMaster University, Hamilton, Ontario, Canada; Institute of Genetics, University of Koln, Germany; Max Planck Institute, Berlin and National Institute of Health, Bethesda, USA as well as Trent Polytechnic, Nottingham, UK) and all of them agreed to take me for a Post-doctoral position to work on different projects and each involved working on *Escherichia coli*, except the last one where a lectureship position was offered. I decided to stay in UK and accepted that position and I believe that it was one of the best decisions I made.

At the Polytechnic the research environment was not that strong and I had to wait about four years before one day I received a letter from Professor Abe Eisenstark, Director of Biological Science Department, University of Missouri, Columbia, USA, telling me that he had a grant for a post-doctoral position and should I wish, he could send me an appointment letter. My Polytechnic director very kindly granted me a sabbatical for 9 months. Abe's team was working on ultraviolet A (UVA) light (then very few laboratories were working in this field). Abe's group had found that although UVA on its own has very little effect on T7 phage inactivation, in the presence of hydrogen peroxide (H_2O_2) a synergistic effect could be observed. Nothing was known about the photochemistry or mechanism of this action. I was assigned this project and discovered that exposure of H_2O_2 by UVA generated superoxide anion ($O_2^{\bullet-}$) and this molecule was responsible for phage inactivation. It was another fortuitous field for me which led me to move to working on free radicals, then not considered as valuable as now.

From this visit I manage to publish three research papers (my own and with Abe) and due to my performance Abe gave me three more opportunities to visit his laboratory which finally gave rise in 1998 to publish a review article on thymineless death in bacteria in *Annual Review of Microbiology*. Furthermore, the discovery of $O_2^{\bullet -}$ generation from the NUV photolysis of H_2O_2 led me to look for other biological chromophores which can be photolysed by Near-UV, and indeed we (my team) showed that a number of amino acids such as phenylalanine, histidine, tryptophan tyrosine as well as β-phenyl pyruvic acid and mandelate can also be photolysed by NUV which can generate a variety of reactive oxygen species (ROS). Meantime one of the team members started working on the synergistic action of UVA light plus 8-methoxypsoralen (PUVA). This twist in my research once more took me to the Genetics Department of Leicester University, this time to collaborate with Professor Barry Holland to extend work on PUVA. There I discovered a protein of 55 kDal which was suggested to be involved in DNA repair in *E. coli* when damaged by PUVA.

While at Nottingham Trent University I also had a number of research collaborations abroad including with the University of Leiden, Holland, and the finding was (published in *Mut Res*) that DNA polymerase-1 in *E. coli* is an inducible enzyme involved in DNA repair; then with the university of Odense in Denmark, with Public Health Laboratory Services, Porton Down, UK, with CNRS at Gif-sur-Yvette in France, and then with scientists at the Institute of Genetics at Moscow. This last collaboration was to study the after-effects of radioactivity fall-out from the accident at the Chernobyl nuclear plant. We showed that a number of radio-resistant bacterial species had emerged there (published in *J Photochem Photobiol B*).

In 2000 I was invited by Professor F. Hanaoka (Fumio) of the University of Osaka, Japan. With another sabbatical from my university I went there for nine months as a Senior Scientist to work on the role of 55 kDal protein in DNA repair. Here, we unexpectedly found that this protein was malate dehydrogenase but what role it was playing in DNA repair could not be answered. Also here I started writing a review on Fanconi anemia (FA) which was triggered by the information that FA cell lines were sensitive to PUVA. *BioEssays* published this work and subsequently I was invited by Landes Bioscience publisher in USA to write a book on FA. Helped by a number of specialists in the field, a book was produced on this children's syndrome.

Then again in 2007, I went to the same University as Senior Scientist and Visiting Professor where I delivered a series of lectures and seminars, but more importantly to extend the work where it was left. Intriguing results were collected and a paper was published in 2012 in which, apart from other information we showed that in fact besides malate dehydrogenase, in *E. coli* there were other enzymes, such as succinate dehydrogenase, and NADH: ubiquinone oxidoreductase playing roles in preventing DNA damage. The conclusion was that the endogenously over-produced ROS in cells were scavenged by one or more of these enzymes and in syndromes such as FA, the removal of ROS by any of these enzymes (possibly due to gene mutation) was incomplete hence the disease.

My other research project, supported by Imperial Chemical Industry of UK, was on genetic manipulation of *Brevibacterium helvolum* to produce high quantity of thymidine to be used in the production of a drug. The paper was published and the researcher obtained an M Phil degree. Ironically the outcome of the result could not be used by the industry. Other publications in Medical hypotheses included improved treatment of burn infection, curing leprosy and treating certain solid tumors.

Since publication of the book on FA a deeper interest developed to publish other medical books and for this I took an early retirement and now spending most of my time in producing these books. Springer published most books and one by CRC Press. The eye-catching books are on: (i) Diabetes (ii) Neurodegenerative Diseases (iii) Obesity and (iv) Diseases of DNA repair.

As Bob had laid the foundation of all my achievements, I consider it incomplete if the readers did not know how my beloved Professor has brought an important twist in my life and hence I wish to present my most cordial gratitude to him by saying that: Bob had not only put my destiny to a successful path from his highly professional academic support and extreme kindness, but today where I am standing specially academically, its seed was watered and fertilized by him. Although academic supports and kindness had also been provided by Abe and Fumio it is not difficult to judge that the kindness showered by Bob had been of superior quality than the latter two professors. Hence in my book on "Neurodegenerative Diseases" in the DEDICATION page, I included those three names that have played major roles in my academic career and showered kindness and affection, giving Bob's name above all others.

Even though I left the Genetics Department in 1972 I kept regular contact with Bob, mostly visiting his house at Knighton Grange Road. At family level he had several ups and downs in his life which he used to share with me until the day when we had our last meeting on a dinner table at an Indian restaurant on London Road, Leicester. One week after, I was shocked to learn that he was admitted to a hospital in Leicester. I visited him with his (step) daughter Naomi. We were informed of the seriousness of his disease and few days later Bob went into a sub-conscious or unconscious state of mind. From there he was removed to a nursing home and I continued my visits, sometimes alone and other times (especially at Christmas time) with Naomi and his second son (now the late) Simon Pritchard. Bob kept on losing recognition as time went by and each time I saw him, as I recalled his utmost kindness and highly friendly manners and the tragic disease he was suffering from, tears came to my eyes that "Bob never deserved this and other hardships he suffered". Here I have to confess and present my appreciation for Naomi and her family who gave their immeasurable love, utmost care, kindness and attention from the very beginning, by making regular visits to the nursing home.

Although Bob has lost his life, he was a jewel in a crown and like a "candle which burns to show light to the people around it and then melts down to eternity". Memories of his dedicated life to science, to politics, to Naomi and her families, his ex-students and Post-doctoral fellows, friends as well as all well-wishers will be staying in our minds and hearts as long as we live.

My Publications with Professor R.H. Pritchard:

Ahmad SI, Barth PT, Pritchard RH. (1968) Properties of a mutant of *Escherichia coli* lacking purine nucleoside phosphorylase. *Biochim Biophys Acta* **161**: 581–583.

Ahmad SI, Pritchard RH. (1969) A map of four genes specifying enzymes involved in catabolism of nucleosides and deoxynucleosides in *Escherichia coli. Molec Gen Genet* **104**: 351–359.

Ahmad SI, Pritchard RH. (1971) A regulatory mutant affecting the synthesis of enzymes involved in the catabolism of nucleosides in *Escherichia coli. Molec Gen Genet* **111**: 77–83.

Ahmad SI, Pritchard RH. (1972) Location of gene specifying cytosine deaminase in *Escherichia coli. Molec Gen Genet* **118**: 323–325.

Ahmad, SI, Pritchard RH. (1973) An operator constitutive mutant affecting the synthesis of two enzymes involved in the catabolism of nucleosides in *E. coli*. *Molec Gen Genet* **124**: 321–328.

Barth PT, Beacham IR, Ahmad SI, Pritchard RH. (1967) Properties of bacterial mutants defective in the catabolism of deoxynucleosides. *Biochem J* **106**: 36–37.

Barth PT, Beacham IR, Ahmad SI, Pritchard RH. (1968) The inducer of deoxynucleoside phos-phorylase and deoxyriboaldolase in *Escherichhia coli*. *Biochim Biophys Acta* **161**: 554–557.

Pritchard RH, Ahmad SI. (1971) Fluorouracil and the isolation of mutants lacking uridine phosphorylase in *Escherichia coli*: location of the gene. *Molec Gen Genet* **111**: 84–88.

12 Professor Robert (Bob) Pritchard (1930–2015): In Memoriam

Ifor R. Beacham*

It was close to half a century ago (1967) that I entered Bob's laboratory as a newly minted graduate in Biochemistry from the Department of Biochemistry, University of Otago, New Zealand. With Bob's passing it is a privilege to reflect on this extraordinary man and his influence.

My mentor at Otago, Dr Mervyn Smith, obtained a studentship for me with one Professor Robert H. Pritchard in the newly established Department of Genetics at Leicester University, and I was encouraged by Professor Edson to "work on nucleic acids". This was at an exciting time in biochemistry, when (in Erwin Chargaff's words) genetics had been "given a chemical education" and microbial genetics was at the forefront of the emerging "new genetics" and genetic manipulation.

The Department at this time was located in an old Law building. Here I remember Bob having a major, animated, tea-time discussion with Jim Shapiro about the evidence (by F. Jacob) for the "promoter"! On one occasion it was necessary to climb through the window in late evening to do something at the bench in a timely fashion! Soon we moved into the Biological Sciences building and matters of access were resolved. Bob had built an exciting Department with wonderful people who always had time

*Institute for Glycomics, Griffith University Gold Coast Campus, QLD 4222, Australia (i.beacham@griffith.edu.au).

for discussion with a naive young graduate; I remember in particular, and with great affection, Drs. Jenny Dee (Physarum; the **true** slime mould!), Susan Hollom (*E. coli*), Clive Roberts (fungi), Brian Wilkins (*E. coli*) and Barry Holland (*E. coli*). There were frequent discussions with Bob, around the blackboard in the laboratory, rather than formal meetings, where, I might add, smoking was not yet banned!

A Problem for a Biochemist

The main interest of the Pritchard laboratory was the control of DNA replication in *E. coli*. New DNA synthesis was typically measured by the incorporation of C^{14} or H^3-thymine. But wild-type *E. coli* does not take up thymine. However, "thymineless" (*thy*) mutants **could** be isolated which would not synthesise thymidylate *de novo* but which could be rescued with high concentration (200 μm) of thymine (otherwise they would not exist, of course); furthermore, secondary mutants could be isolated which only required low concentrations (20 μm) of thymine — not unreasonably termed *tlr* (thymine low requirement).

How can thymine be taken up by *thy* mutants, but not by the wild type, and what was the basis for mutation to low requirement? Bob decided that a biochemist was required to solve this problem! This was not easy, but it was resolved in the end though not entirely due to our own efforts.[1–5] Furthermore, as Bob hypothesised, the external thymine concentration influenced the rate of a replication fork (with profound consequences, as discussed elsewhere).[6]

Bob's Mentorship

I should begin with Bob's undergraduate genetics lectures which I attended. These were quite brilliant, incisively outlining the basics of recombination and complementation. Of course, Bob was already "living history" with his work showing directly that the gene, in *Aspergillus nidulans*, had linear dimensions (published concomitantly with the Benzer studies using phage).

His lectures later informed my own genetics lectures; the common misuse (or, to be charitable, adaptation) of the term "complementation" still rankles with an old biochemist-turned-geneticist!

Bob was a friendly and close mentor; discussions at the blackboard would often veer onto other topics and his sharp mind was a delight. When it came to my PhD 'viva', with internal examiner Professor Sir Hans Kornberg (Chair of the adjacent Department of Biochemistry), Bob was present as support. I am not sure if this was usual!

Bob taught me very early on that asking *the right questions* was the important beginning of an investigation. Of course it was more than the right question; it was the appropriate question also — as Medawar famously remarked, "science is the art of the possible".

I recall several trips to conferences which Bob encouraged and facilitated. One such was to Edinburgh to meet with the bacterial genetics group there — he drove us all the way with frequent discussion *en route* on all sorts of topics both political and scientific. Present at the meeting were Frank Stahl, Julian Gross and others. Another trip was to the 3rd Lunteren Symposium in 1970 in The Netherlands, which centred on replicons and episomes. William Hayes and Bob gave outstanding presentations. My own minor contribution was nervously presented and I remember thinking that no one would be interested in how the thymine got there!

His influence has echoed down the decades, since the Leicester years, with genetics, in its various forms, forming the basis of my own research and informing my own undergraduate teaching and graduate mentoring. He also, unwittingly, taught me the importance of friendship and open collaboration in science and as a consequence I have always been close to my own post-graduate students and have maintained those friendships over many decades.

"The road goes ever on and on, down from the door where it began"

(JRR Tolkien)

Of course this literature was very popular around the time of my PhD tenure! The quote continues *"...and whither then, I cannot say"*. Well, nearly

50 years later, that 'door' to science, and bacterial genetics in particular, clearly reverberated into my future.

Bob once told me that if he left *E. coli* he would leave biology, and he was true to his word. In my later years, in contrast, I felt that I would like to explore new horizons, specifically other bacteria, notwithstanding that I (and many others presumably) still claim *E. coli* to be the greatest model organism ever for studying basic questions in molecular biology (of course!). I chose a pathogen, *Burkholderia pseudomallei*, a bacterium that, at the time, had not been worked on very much at the molecular level, but caused a potentially fatal disease, melioidosis; furthermore, it was mainly endemic in Asia and Northern Australia and also over-represented in our Australian indigenous population. Once again we brought the genetic manipulation guns to bear, so to speak. I like to think that Bob, the humanist, approved of this project.

But what a wonderful start by a great thinker and scientist. The road certainly did go ever on and Bob started me on that great journey.

What memories.
Thank you.

References

1. Barth PT, Beacham IR, Ahmad SI, Pritchard RH. (1968) Properties of bacterial mutants defective in the catabolism of deoxynucleosides. *Biochem J* **106**: 36–37.
2. Barth PT, Beacham IR, Ahmad SI, Pritchard RH. (1968) The inducer of the deoxynucleoside phosphorylases and deoxyribaldolase in *Escherichia coli. Biochim Biophys Acta* **161**: 554–557.
3. Beacham IR, Barth PT, Pritchard RH. (1968) Constitutivity of thymidine phosphorylase in deoxyribaldolase negative strains: Dependence on thymidine requirement and concentration. *Biochim Biophys Acta* **166**: 589–592.
4. Beacham IR, Eisenstark A, Barth PT, Pritchard RH. (1968) Deoxynucleoside sensitive mutants of *Salmonella typhimurium. Mol Gen Genet* **102**: 112–127.

5. Beacham IR, Pritchard RH. (1971) The role of deoxynucleoside phosphorylases in the degradation of deoxynucleosides in thymine-requiring mutants of *E. coli*. *Mol Gen Genet* **110**: 289–298.
6. Beacham IR, Beacham K, Zaritsky A, Pritchard RH. (1971) Intracellular thymidine triphosphate concentrations in wild type and in thymine requiring mutants of *E. coli* 15 and K12. *J Mol Biol* **60**: 75–86.

13 Personal Recollections of an Exciting Scientific Period (1969–1971 and Beyond): A Tribute to Bob

Arieh Zaritsky*

Introduction

This is devoted to Bob centering on my experience and accomplishments that stemmed directly from his high level of scientific ingenuity and personality. Much of it has become known to me during the three formative years under his superb guidance and many years of further interactions. The description of my experience under Bob's supervision and its 'after effects' is meant to clarify additional features of this outstanding scientist and humanist. Personal notes are included in passing. I try to avoid repetition and complement the obituary so well written by **Peter A. Meacock**.[1]

During six years (1962–1968) of academic studies at the Hebrew University of Jerusalem (HUJI), I graduated (MSc, with distinction but no publication) in Bacterial Genetics supervised by **Amiram Ronen**, finished four full years of Pre-Medical studies and attended several courses in Mathematics, my ever-lasting love, likely stemming of some familial genetic elements.[2] After "tasting the Apple of Knowledge" (basic math

*Faculty of Natural Sciences, Ben-Gurion University of the Negev, Be'er-Sheva, Israel (ariehzar@gmail.com, http://ariehz.weebly.com/).

and scientific research), my interest in becoming an MD faded away. There was no sign or wish to leave my homeland (with a British "Palestine-Eretz Israel" birth certificate of 1942) until my application for PhD studies in the field of Prokaryotic DNA was rejected by **Erela Ephrati-Elizur** and several Virologists. I was accepted by **Shmuel Razin**, but his field of research, bio-membranes, did not appeal to me, so I started to look for studies abroad.

The Choice of Bob at Leicester, the United Kingdom

Shortly after the Six-Days War, reading Churchill's six volumes about WWII (bought as a gift for my father's 60th birthday) intrigued me: England's situation during "The Gathering Storm"[3] resembled Israel's challenges by all Arab surrounding countries in both 1947–9 and 1967. **Giora Simchen**, just recruited as a young Genetics lecturer at the HUJI, arrived from Birmingham UK and was exceedingly helpful in consulting and composing letters approaching British scientists in the field. Of the seven letters sent, six responses arrived, two of which were positive: from **Robert Hugh (Bob) Pritchard** and from **William (Bill) Hayes**, the former had been appointed in 1964 (at the young age of 34!) as the Establishing Chair of the new Leicester University's Genetics Department — the latter just moved his MRC Bacterial Genetics Unit from Hammersmith Hospital, London to the University of Edinburgh. At the time, it seemed to be a 'catch-22' situation, but according to my 'tradition', I let each know about the other's option. Bill's response (May 9, 1968) was swift, honest and generous to Bob (*...I know Pritchard well and you could hardly do better...*), thus resolving my dilemma! I was unaware that consequent to Bob's training at Bill's unit (1959–1963) they were very fond of each other until the three of us met in the SGM's Annual Meeting in London, Spring, 1969.

I was fortunate, again: Giora had heard about the new 'star' geneticist who was among the youngest department Chairmen at the time, and highly recommended and encouraged me to go there. In both arenas, Science and Society, Bob followed the footsteps of his admirable mentor, the leading Geneticist **Guido Pontecorvo** (*aka* Ponte) at Glasgow. For example, in 1955, when promoted to the newly created Chair of Genetics there and elected

Fellow of the Royal Society, *he circulated a note saying that henceforth the head of department should be known as "Ponte" — meaning of course no change.*[4] Similarly, nobody surrounding Bob has ever dreamt naming him "Professor Pritchard", as accustomed by most professors in those 'good, old days'; both were *anything but pompous*.

The Leicester department, designed by Bob 10 years later, survived in temporary accommodation for a while before moving to the new purpose-built Adrian Biology Building in 1968, where they shared the first floors with **Hans Kornberg**'s Biochemistry department. The new, considerably expanded department was run as a 'commune': a single kitchen that served all and saved on redundant manpower, a joint meeting room used to exchange opinions and advance ideas about Science and Society alike, grant sharing with temporarily unfortunate colleagues, who ran their own laboratories absolutely independently but inspired by intense discussions that clarified their specific fields. This socialistic spirit persisted for at least half a century, as evident during the 50th anniversary shortly before Bob passed away.[5] Each of these two eminent scientists (Ponte and Bob) founded a Genetics department at his respective university and served it during 23 years, 1945–1968 (GP) and 1965–1988 (RHP), a coincidence that may have evolved in Bob's mind, retiring early at age 58. Similarly, their departments *...had distinctly international flavor... there was a stream of visitors, PhD students and postdoctoral scientists, from Europe, America and Asia....*[6] Both professors personally taught the first course in Genetics to freshman biology students during all those years. Both were instrumental in modernizing their departments in their mid-term, Ponte to Molecular Biology, Bob to Genetic Engineering and Human Genetics at the newly-established Medical School. Glasgow, however, was more rewarding than Leicester: Ponte was awarded an honorary degree and a devoted building named after him; wouldn't it be appropriate for the latter to imitate Glasgow?

Ponte's heritage of an old Italian (nonobservant) Jewish family likely affected Bob's extreme secular outlook, occasionally remarking that one of his own grandmothers might had been Jewish. Not that it matters, but interesting it is, no doubt. Both passed a period of exile, though totally different: whereas Ponte fled persecution by Nazi-inspired racial policies, Bob spent several happy years of his youth in "the country", escaping the

The mythological teacher Ponte and his student Bob in their later years; time and venue unknown. (*From Bob's personal photos collection, given by Naomi Matthews (Pritchard).*)

Blitzkrieg over London. Both persons were *...non-conformist*(s), *politically left of centre but admirer*(s) *of British liberalism....*[6]

Like Ponte, Bob started his career as a Botanist, investigating fungal genetics. In 1945, Glasgow's was the only British department dealing with genetics of microorganisms. Ponte was inspired by his mentor **Herman J Muller'***s understanding of the gene as a continuum*, to be tested by intragenic recombination. Bacterial Genetics was just discovered by the ice-breaking fluctuation test,[7] and the Phage Group scientists led by **Max LH Delbrück** galloped to confirm Muller's hypothesis by intragenic linkage map in *rII* mutants of the T4 bacteriophage.[8] Bob's series of experiments with adenine-requiring mutants of *Aspergillus nidulans*,[9] a system developed by his mentor,[10] was simultaneously published; **Seymour Benzer'**s studies are more famous due to the much higher resolution power of the phage system as well as by the introduction of the term "cistron". Ponte's understanding, like that of Delbruck's team, was advanced *purely by genetic analysis, with no input from biochemistry*. His devoted follower diverged: Bob believed in the 1960s that understanding the mechanism governing DNA replication would progress by deciphering precursors' metabolism, and faster so in bacterial systems. Fortunately, Bob was hired by Bill at a Cold

Bob (L) while AZ (R) visited Leicester during an EMBO fellowship at CL Woldringh's laboratory (Amsterdam, 1977).

Spring Harbor meeting due to his notable work under Ponte's mentorship (personal communication to **Paul Broda**); there, at the Hammersmith MRC Unit he became a Bacterial, Molecular Geneticist. With the same sharp analytical mind Bob advanced details of those pathways in *Escherichia coli*, in parallel with deep understanding of Cell Physiology and Genetics. The discoveries made by his students under his instructive leadership allowed Bob to integrate the fields, elucidating that deep understanding of a cell needs a thorough knowledge of — even feeling for — its behavior as a whole system (e.g. Ref. 11).

Reaching Leicester

There was only a 'small' item left: how will Bob support my expenses? After several refusals to scholarship applications, my parents proposed to cover the costs, but being 26 years old and after two years of employment as a

Bob Pritchard (L) and Kurt Nordström (R) at the EMBO Workshop in Segovia, Spain, 1987. (*Photographed by Alfonso Jiménez-Sánchez.*)

Teaching Assistant at the HUJI's Genetics department, I was not receptive to the idea. So, in September 1968 I stopped MD studies and continued to work as an Assistant at the Cancer Research department, which culminated in my first peer-reviewed, co-authored article.[12] Bob's application for my EMBO Long-term Fellowship was finally approved in November, apparently by Bill, one of the Founding EMBO members, and I took off to London late in December. One should keep in mind that there was no fax let alone email in those days and a response to each letter took at least 10 days to arrive...

Expressing my sincere desire to start work, I flew to London then travelled by train to Leicester a week before the date my scholarship would commence (January 1, 1969) — totally ignorant of the Christmas vacation... The Adrian Building was empty, except for **Barry Holland**, who challenged me with accusations about Israel's "unjustified, disproportionate violence" ('blood for Matzah?'[13]..); he was nevertheless very helpful, then and whenever needed afterwards. Much help was extended to newcomers by many members, particularly **Susan Hollom** (later **Wilkins**) can be commended. Bob hosted me and we had pleasant conversations about various subjects, despite my broken English. For example, Bob accepted foreign students because (freely cited) *"those who take the trouble to come from afar prove to be highly serious and motivated"*. Indeed, in addition to half a dozen UK graduate students, Bob instructed those days (1965–1985) also an Indian (**Shamim I. Ahmad**), a Sri-Lankan (**Horadigala Gamage Nandadasa**, *aka* Das), two established Israeli scientists on Sabbatical (**Robert F. Rosenberger** [**Bob R.**] and **Ezra Yagil**) and several Spanish trainees. Warm friendships arose among them/us all, and several cooperation modes have emerged. For example, almost 20 years later, together with Das, we were awarded a US-AID research grant for our endeavors to eradicate Malaria through Biological Control of Mosquitoes using transgenic organisms, which culminated in a cooperative research with Ezra in the 2000s.[14,15] Several articles with Bob R, one (my second) in *Nature*,[16] stemmed from a research team that I organized during the 1970s and 1980s together with **Conrad L. Woldringh**. A student of **Elena Guzmán** was recently awarded a short-term EMBO fellowship to investigate cell division in Be'er-Sheva together, to be published soon.

With a then-plentiful fellowship (*ca* 1,400 £/year; it was still the British £sd era...), first thing was to purchase a car (yellow Austin Mini-Minor of 1962) and spikes sprint-running shoes, items that I could not afford in Israel, heavily taxed those meagre days. The shoes enhanced my frequent jogging in Oadby's track, and the Mini served my frequent visits to London and roaming the UK during the 3-years' studies period. The Friday's weekend paper 'Ma'ariv', reaching Leicester *on* Saturday (how efficient were the post services those days!) kept me in pace with the events in Israel. No internet of course, and no trust in British magazines... Otherwise, it was 18 hours/day work. And indeed, highly rewarding it was!

Choosing a Project and Defining the Working Hypothesis

With the back-to-back manuscripts of articles just published by **Charles E. Helmstetter** and **Stephen Cooper** (1968),[17,18] I was sent to learn them and to the library to pick a project for myself. At that stage, I was completely ignorant in Bacterial Physiology and the Cell Cycle and hence came up with strange, irrelevant ideas. Here came the subtle leadership of Bob to play a major role in shaping my future: after several discussions, it seemed to me that I decided by myself to go after the rate of chromosome replication. A collection of literature-recorded results totally disagreed with Cooper & Helmstetter's "BCD" model (see in Ref. 19), thus misled interpretations of excellent scientists. Bob's out-of-the-box but facts-based, crystal-clear thinking mode directed him to an explanation that was consistent with all these apparently strange observations, the common denominator of which is that they had been obtained in thymine auxotrophic strains. And since Bob was deeply involved with studies on the pathways of DNA building blocks (see e.g. Ref. 11), this common feature led him to hypothesize that the rate of chromosome replication in *thyA* mutants depends on the concentration of thymine [T] supplied in the growth medium.

The general picture clarified when the experimentally-derived BCD model[19] was combined with the 10 years-old measurements by **Niels-Ole Kjeldgaard, Ole Maalϕe** and **Moselio Schaechter**[20,21] of cell size and macromolecular composition under varying growth rates and during transitions. Together with the definitions of cell age distribution[22] and steady-state growth,[23] average cell size M[24,25] and DNA content G[18] were found to depend on four parameters, all mean values: culture and cell doubling time τ, chromosome replication time C, time from replication-termination to cell division D, and cell's mass per *oriC* at initiation M_i, the latter three being roughly constant (40 min for C, 20 min for D, and strain-dependent M_i) at a constant temperature (37°C). Thus, simultaneous determinations of M and G of *thyA* mutants growing 'normally' under steady-state conditions in glucose minimal medium supplemented with various [T]s would allow calculating the values of C and D as a function of [T]. Identical mass growth rate (τ^{-1}) under varied replication rates (C^{-1}) would complement the idea[17,18] of dissociation between the two. Strain 15T$^-$ of *Escherichia coli* was picked for this study[26] and the results were soon afterwards confirmed in strain K-12 CR34.[27,28] Selecting these strains proved later to be fortunate, when cell shape and dimensions were considered (e.g. Refs. 19, 29; and see below).

Advancing the Project: Solving Difficulties and Conclusions

Our attention was quickly drawn to the fact that the inter-division time τ_d of cells grown in glucose medium (with $\tau \leq ca$ 50 min) was longer than the mass doubling time τ_m, thus the culture does not reach steady-state but rather grows "normally",[30] and that τ_d was longer at lower [T]'s without observed change in τ_m.[31] This faulty division frequency directed our suspicion toward the *D* period, the molecular basis of which is still enigmatic today. This project was thus saved by calculating the ratio *G*/*M*, so-called "DNA concentration", to cancel *D* from the analysis,[26] as follows: $G/M = (\tau/C\ln 2)[2^{(C+D)/\tau} - 2^{D/\tau}]/M_i\, 2^{(C+D)/\tau} = [\tau/(M_i\, C\ln 2)](1-2^{-C/\tau})$. And since the ratio C/τ had been defined as the number of 'replication positions' *n*,[32] $G/M = [M_i n \ln 2)]^{-1}(1-2^{-n})$ is a function of *n* alone. Relative values of *C* can thus be derived from measuring τ and relative values of *G*/*M* irrespective of the values of *D* and M_i. To discern small differences in DNA concentration, [^{14}C]-labelled thymine was exploited. The specific radioactivity (Curies per mole) had to be kept absolutely constant within an experiment comparing *G*/*M* values; this was achieved by appropriately diluting the growth medium containing high ^{14}C-labeled [T] in identical medium without any added thymine. In retrospect, this technical, simple trick was probably not taken by several scientists, who were unable to repeat our experiments (personal communications).

Another measure that depends on *n* alone is the increment of DNA after release from a specific blocking of mass growth by e.g. removal of required amino acid $\Delta G = [(2^n\, n \ln 2)/(2^n - 1)] - 1$; at a given τ, *C* thus uniquely defines ΔG. A series of such measurements in cultures growing in an identical medium but supplied with different [T]'s provided independent estimates of *C*, and in absolute time values,[26] assuming that M_i is not affected and remain constant in a given strain and temperature.

Three additional, direct approaches that are independent of *D* were performed *in vivo*, both affected by a sudden change in [T], so-called step-(-up or -down): (a) the change of DNA synthesis-rate, and the time it takes to reach (b) the new steady-state value of *G*/*M* or (c) the plateau level of ΔG.[26,31,33] All five methods, each dependent on a different set of assumptions, converged to agree with Bob's hypothesis, concluding that the rate of chromosome replication in *thyA* mutants can vary over a range of 2–3-fold by modulating [T] without significantly affecting τ or cell viability.[26] This conclusion was soon afterwards confirmed by biochemical methods[34] and in thymine auxotrophic *Bacillus subtilis* by Erela Ephrati-Elizur at the HUJI (Ref. 35; and see Introduction above). These methods have since been refined, expanded and exploited by Bob's disciples (e.g. Refs. 33, 36–39) and by others, to demonstrate various phenomena, as described elsewhere and partially summarized in Ref. 40.

Long-term Insights

An additional method developed to this end was based on the findings of Bob himself[41] and extended by his first student **Peter T. Barth**:[42] the degree by which the rate of DNA synthesis increases following a specific inhibition for one mass doubling time, so-called "Rate Stimulation Factor" (RSF; Ref. 43). This measure is also a function of *n* alone, $RSF = [2^{(n+1)-1}]/(2^n-1)$, but its sensitivity to *n* is quite low. Nevertheless, an offshoot of this investigation (Figs. 6 and 7 in Ref. 43) ...*suggested the existence of a minimal possible distance between two successive active forks along the chromosome*... And that was stated, mind you, 40 years ago! Despite hot arguments from my colleagues, then (**John Collins**) and later (**Conrad L. Woldringh**), that the chromosome is long enough to accommodate and manage many replisomes, I "boldly" dared to bluntly state this conclusion, likely because Bob taught me to stick to one's results and to express one's own mind independently of others... Only *ca* 30 years later, when increased initiations by excess *dnaA* expression was demonstrated to result in "*replication fork collapse*" and "*DNA double-strand breaks*", I returned to my original suggestion — this time coining the term "Eclipse".[44] The highly rewarding bonus is that this thought can explain the puzzling phenomenon of abnormal cell divisions mentioned above,[31] to be dealt with in the future. Bob and Peter deserved co-authorship on the original article[43] that described RSF; their names were not included only because they firmly refused. Concordant with Bob's generosity, they requested me to leave their names out because the recorded results were obtained by me; in retrospect, I feel guilty for not insisting their co-authorship.

During my first two years at Leicester I have not used the microscope while performing the various experiments, assuming that the larger cell sizes resulting of longer *C* values are accommodated in the length dimension as happens under thymine starvation (at $[T] = 0$). That is why the marked increase in cell width did surprise me and Bob, to say the least. The first stunning observations of thickened, then branched cells[31,44–47] were obtained while composing my Thesis[48] hence not reported in it. This basic observation, that diameter of the cylindrical cell rises not only by nutritional up-shift but also by replication-rate down-step,[31,44–47] haunted me for several years. It has apparently haunted Bob as well.[49] Our seemingly separate routes converged again, experimentally, by my encounter with

Conrad Woldringh (as described in Ref. 19). Serendipitously, this encounter stemmed of sleepless nights during which I attempted to combine and integrate the dimensional changes with the kinetics parameters, *i.e.* spatial and temporal aspects of cell duplication. It was a 3am "eureka" while guarding the camp as a reservist soldier during the 1973 war; the idea hit me like a strike, and was quickly written for publication.[50] This encounter was another luck-strike, at least for me, resulting in an ongoing cooperation with Conrad for over four decades.[19]

Among the many studies jointly performed in Amsterdam (mostly with Conrad, some with the computer engineer **Norbert O.E. Vischer**), which culminated in over 20 coauthored, peer-reviewed articles (see in http://ariehz.weebly.com/articles.html), was developing the Cell Cycle Simulation (CCSim) computer program (https://sils.fnwi.uva.nl/bcb/cellcycle/), recently described in Ref. 51. This instructive program was instrumental in supporting the Eclipse idea[43,44] (and watch the video at Ref. 51 or Ref. 52). The buds of this insight were germinated at Leicester some 45 years ago, supervised by Bob. Moreover, the series of Workshops on the Prokaryotic Cell Cycle, financed by the EMBO (e.g. the poster to the 2nd http://ariehz.weebly.com/uploads/2/9/6/1/29618953/emboworkshop1984.pdf), was initiated by Conrad and me as a conceptual and spiritual follow-up to the 1978 Plasmid Replication Workshop organized by Bob and his good friend, the late **Kurt Nordström** (see below their joint picture in the 3rd of the series at Segovia, Spain).

Further Effects and Implications

The success in my Bob's experience was swift and dramatic: the PhD[48] was bestowed in July, 1971, merely 2.5 years after my arrival at Leicester, paid by scholarships from EMBO (1969) and the HUJI's Friends (1970), and as an SRC employee (1971). Having an article in *Nature*,[26] two in *J Mol Biol*[27,28] and one in prep for *J Bacteriol*,[31] I was hired by the then developing new "Negev University" (later named after Israel's Founding Father and first Prime Minister **David Ben-Gurion**) *before* leaving for a 1-year post-doctoral training — a rare privilege for a young scientist. My 2nd long-term EMBO Fellowship was awarded to spend a year (1972) in the laboratory of Ole Maalϕe, the founder of the Bacterial Physiology field and of the prestigious

Copenhagen's University Institute of Microbiology, so-called Copenhagen School.[20,21,53] Ole knew Bob from Bob's visit there as a Rask-Ørsted Fellow (1954–5), upon the breakthrough of Bob's (see Ref. 9 above).

The idea that attracted Ole to immediately accept my application combined mass growth with DNA replication, a natural outcome of my PhD Thesis. It was far-fetched but displayed my understanding of what was going on in both arenas — to Bob's credit, again. Following growth in numerous experiments, a strange trend was observed namely, rate of mass increase (τ_m^{-1}) was systematically *faster* at *lower* [T] (*longer C*). The hypothesis that I entertained (see Ref. 54 and references therein) was related to the fact that most operons encoding rRNAs are placed near *oriC*. Thus, their relative gene dosage[36] rises at slower replication rate resulting in higher level of ribosomes that will shorten τ_m which, in turn, will increase the frequency of initiations hence raise still further the dosage of these rRNA-encoding genes. This vicious cycle could only be imagined at those old days, when our ideas about growth regulation were rather primitive.

As mentioned above, long-term friendships, joint grants and articles, out-of-the-box innovations and many other 'bonuses' all through my scientific life stemmed from these three formative years of my career and personality.

The Innovative, Humble and Bold Scientist

Bob's facts-dependent, unbiased scientific mind is an example for all of us, his disciples. He was modest and humble: "I am a teacher" was the answer to a "what is your profession?" query by those who did not know him. Nevertheless, Bob was aware of his innovative power: in the 1970's Lunteren Conference, I vividly remember the enthusiasm by which he tried to convince participants that chromosome replication is bi-directional. Several articles were published in 1971–2, demonstrating this to be true — none mentioned Bob's contribution... Moreover, not once his ideas were scooped, but response to our queries about such cases was (freely cited) "innovations arise repeatedly by some..." Indeed, innovative he was, with intragenic recombination[9] and mechanism of recombination in fungi,[55] with the so-called "pre-mature" initiation of replication[41] and negative regulation of this process,[25] with metabolism of DNA-precursors,[11] with integrative suppression of *oriC*-inactivation by plasmids,[56] with control of mitosis in eukaryotes[57] etc. [e.g. Refs. 36–38], to be told by those who were involved with him in these and other original findings. Conrad reminds me that during our drive to the 1997 EMBO Workshop in Chorin (Berlin), Bob, who travelled with us, emphasized that the primary trigger of cell division could not be a protein because that would require still an earlier protein, *ala* "protein-cannot-make-protein paradox".[58] It should therefore be some physical signal — an elusive idea that we have been entertaining for a long time but is hard to demonstrate.[19,29]

Bob was among the few who publicly and boldly refuted[1,59] the recombinant DNA moratorium[60] declared in 1975 by several famous scientists in the field, some of whom were Nobel Laureates. Very early, he predicted the current problems with the indiscriminate use of antibiotics, proposing establishment of an "antibiotic forecast" system (personal communication; and see Peter Barth, above).

It is strange (to say the least) that Bob was not awarded an FRS status. His admirers like us believe that his liberal views in conservative England were the major reason. Indeed, an extreme liberal he has been during all his life. Bob was involved in real-life politics, joining the Liberal-Democratic party, first in the Leicester council and later attempting to be elected to the Parliament; the latter was not successful though. His view that drugs should be freely available (albeit after educating the youth to understand the consequences of using them) did not help the decisions in as conservative institute as the Royal Society of London.

Bob did not hesitate and succeeded to recruit another Professor to the department, **RFS Sir Alec Jeffreys,**[61,62] who pioneered DNA fingerprinting, thus transforming police investigations, letting him free of the usual tasks of a professor. Wishing to retire, he recruited a third professor, an attempt that was not successful.

Personal Life and Interactions

At the time, Bob was a single-parental father raising two boys (**John** and **Simon**), later joined by **Naomi**, adopted when he married **Susan**. He devoted early evenings to getting them to bed and quite often returned to the laboratory. There were many a time that we spent long nights following results flowing from the scintillation counter or cell (Coulter) counter and discussing the predictions and the outcomes — a highly inspirational experience for a graduate student. An experiment that required frequent sampling was performed jointly, harmoniously and impeccably to the delight of both of us. Losing both his boys before his terminal illness was a terrible tragedy. Luckily, Bob has had Naomi, who graduated as a lawyer, and her young family to help him during the last tortuous leg of his life.

Bob was liked and appreciated by all who knew him, even when he expressed diametrically opposing opinions to theirs. Personally, I was shocked by his declaration that (freely cited) "Britain mistook in planting Israel in the Middle-East" and that he will "endeavor to correct this mistake". Never mind that Israel was established *despite* British refusal to let it free from the League of Nations' Mandate and after successfully fighting against the Empire, but remembering the inexcusable blockade against arrival of Jewish refugees to their homeland, encamping them behind barbed wires in Cyprus after their terrible traumas in Nazi Germany's bitter Holocaust 'experience'!

Another interesting words-exchange occurred when we saw an ultra-Orthodox family while sitting on a coffee-house balcony in the old city of Jerusalem. "How disgusting", Bob said, "look how they educate their kids to continue ultra-Orthodoxy". To which I responded by an instructive, rhetoric question that remained unanswered: "Don't you do the same, educating your children to be atheists, perpetuating your lifestyle?"

To sum up, I must admit to have tremendously enjoyed life in the UK despite encountering spells of racism, particularly against "colored" immigrants. The centuries-long British colonialism around the world has left one important asset to the countries this Empire ruled: retaining the democratic values inherited from the sometimes vicious rulers. Pity that its lifestyle in the first decades after WWII did not prevail 'at home'; in my experience, similar lifestyle still exists in New Zealand — is it the result of isolation from the rest of the world, as the British Isles were for centuries?

References

1. Meacock PA. (2016) Bob Pritchard: Eminent scientist and remarkable individual. This Compendium, pp 8–11.
2. Parikh CA. (1991) *The Unreal Life of Oscar Zariski*. Academic Press, Inc. San Diego-London, 264 pp; later 2009 published by Springer-Verlag New York, 194 pp. Softcover ISBN: 978-0-387-09429-8 (DOI of e-ISBN: 978-0-387-9430-4).
3. Churchill WS. (1948) The Gathering Storm. Vol. 1 in *The Second World War*. Houghton Mifflin Co, NY. ISBN 0-395-41055-X. (Hebrew translation, 1965).
4. Cohen BL. (2000) Guido Pontecorvo ("Ponte"), 1907–1999. *Genetics* **154**: 497–501.
5. Leicester University's Genetics department Pictures (first two) http://ariehz.weebly.com/lab-picts.html
6. Siddiqi O. (2002) Guido Pontecorvo 29 November 1907–25 September 1999. *Biographical Memoirs of Fellows of the Royal Society* **48**: 375–390. http://www.jstor.org/stable/3650267
7. Luria SE, Delbrück M. (1943) Mutations of bacteria from virus sensitivity to virus resistance. *Genetics* **28**: 491–511.
8. Benzer S. (1955) Fine structure of a genetic region in bacteriophage. *Proc Natl Acad Sci USA* **41**: 344–354.
9. Pritchard RH. (1955). The linear arrangement of a series of alleles of *Aspergillus nidulans*. *Heredity* **9**: 343–371.
10. Pontecorvo G. (1953) The genetics of *Aspergillus nidulans*. *Adv Genet* **5**: 141–238.
11. Pritchard RH. (1974) On the growth and form of a bacterial cell. *Phil Trans R Soc London* Ser B **267**: 303–336.
12. Patinkin D, Zaritsky A, Doljansky F. (1970) A study of surface ionogenic groups of chick embryo cell transformed by Rous Sarcoma Virus. *Cancer Res* **30**: 498–505.
13. ADL: Blood Libel: a dales, incendiary calim against Jews http://www.adl.org/anti-semitism/united-states/c/what-is-the-blood-libel.html?referrer=https://www.google.co.il/#.Vl6wNXYrLmE
14. Melnikov O *et al.* (2009) Site-specific recombination in the cyanobacterium *Anabaena* sp. Strain PCC 7120 catalyzed by the integrase of coliphage HK022. *J Bacteriol* **191**: 4458–4464.
15. Zaritsky A *et al.* (2010) Transgenic organisms expressing genes from *Bacillus thuringiensis* to combat insect pests. *Bioeng Bugs* **1**: 341–344.
16. Rosenberger RF, Grover NB, Zaritsky A, Woldringh CL. (1978) Control of microbial surface growth by density. *Nature (Lond)* **271**: 244–245.

17. Helmstetter CE, Cooper S. (1968) DNA synthesis during the division cycle of rapidly growing *E. coli* B/r. *J Mol Biol* **31**: 507–518.
18. Cooper S, Helmstetter CE. (1968) Chromosome replication and the division cycle *of E. coli* B/r. *J Mol Biol* **31**: 519–540.
19. Zaritsky A, Woldringh CL. (2015) Chromosome replication, cell growth, division and shape: a personal perspective. *Frontiers in Microbiology* **6**: 756.
20. Schaechter M, Maalφe O, Kjeldgaard NO. (1958) Dependency on medium and temperature of cell size and chemical composition during balanced growth of *Salmonella typhimurium. J Gen Microbiol* **19**: 592–606.
21. Kjeldgaard NO, Schaechter M, Maalφe O. (1958) The transition between different physiological states during balanced growth of *Salmonella typhimurium. J Gen Microbiol* **19**: 607–616.
22. Powell EO. (1956) Growth rate and generation time in bacteria with special reference to continuous culture. *J Gen Microbiol* **15**: 492–511.
23. Campbell A. (1957) Synchronization of cell division. *Bacteriol Rev* **21**: 263–272.
24. Donachie WD. (1968) Relationships between cell size and time of initiation of DNA replication. *Nature* **219**: 1077–1079.
25. Pritchard RH, Barth PT, Collins J. (1969) Control of DNA synthesis in bacteria. *Microbial Growth. Symp Soc Gen Microbiol* **19**: 263–297.
26. Pritchard RH, Zaritsky A. (1970) Effect of thymine concentration on the replication velocity of DNA in a thymineless mutant of *Escherichia coli. Nature (Lond)* **226**: 126–131.
27. Beacham IR, Beacham K, Zaritsky A, Pritchard RH. (1971) Intracellular thymidine triphosphate concentrations in wild type and in thymine requiring mutants of *Escherichia coli* 15 and K12. *J Mol Biol* **60**: 75–86.
28. Zaritsky A, Pritchard RH. (1971) Replication time of the chromosome in thymineless mutants of *Escherichia coli. J Mol Biol* **60**: 65–74.
29. Zaritsky A. (2015) Cell shape homeostasis in *Escherichia coli* is driven by growth, division and nucleoid complexity. *Biophys J* **109**: 178–181.
30. Fishov I , Grover NB, Zaritsky A. (1995) On microbial states of growth. *Mol Microbiol* **15**: 789–794.
31. Zaritsky A, Pritchard RH. (1973) Changes in cell size and shape associated with changes in the replication time of the chromosome of *Escherichia coli. J Bacteriol* **114**: 824–837.
32. Sueoka N, Yoshikawa H. (1965) The chromosome of *Bacillus subtilis*. I. Theory of marker frequency analysis. *Genetics* **52**: 747–757.
33. Bremer H, Churchward G. (1977) Deoxyribonucleic acid synthesis after inhibition of initiation of rounds of replication in *Escherichia coli* B/r. *J Bacteriol* **130**: 692–697.

34. Manor H, Deutscher MP, Littauer UZ. (1971) Rates of DNA chain growth in *Escheriscia coli*. *J Mol Biol* **61**: 503–524.
35. Ephrati-Elizur E, Borenstein S. (1971) Velocity of chromosome replication in thymine-requiring and independent strains of *Bacillus subtilis*. *J Bacteriol* **106**: 58–64.
36. Chandler MG, Pritchard RH. (1975) The effect of gene concentration and relative gene dosage on gene output in *Escherichia coli*. *Mol Gen Genet* **138**: 127–141.
37. Meacock PA, Pritchard RH. (1975) Relationship between chromosome replication and cell division in a thymineless mutant of *Escherichia coli* B/r. *J Bacteriol* **122**: 931–942.
38. Pritchard RH, Chandler MG, Collins J. (1975) Independence of F replication and chromosome replication in *Escherichia coli*. *Mol Gen Genet* **138**: 143–155.
39. Churchward G, Bremer H. (1977) Determination of deoxyribonucleic acid replication time in exponentially growing *Escherichia coli* B/r. *J Bacteriol* **130**: 1207–1213.
40. Zaritsky A, Woldringh CL, Einav M, Alexeeva S. (2006) Use of thymine limitation and thymine starvation to study bacterial physiology and cytology. *J Bacteriol* **188**: 1667–1679.
41. Pritchard RH, Lark KG. (1964) Induction of replication by thymine starvation at the chromosome origin in *E. coli*. *J Mol Biol* **9**: 288–307.
42. Barth PT. (1968) Genetic and Biochemical Studies on the Control of DNA Replication in Bacteria. Ph.D. Thesis, Leicester University, England, 177 pp.
43. Zaritsky A. (1975a) Rate stimulation of DNA synthesis after inhibition. *J Bacteriol* **122**: 841–846.
44. Zaritsky A, Vischer N, Rabinovitch A. (2007) Changes of initiation mass and cell dimensions by the 'eclipse'. *Mol Microbiol* **63**: 15–21.
45. Zaritsky A. (1977) Branching of fast-growing *Escherichia coli* 15T$^-$ at low thymine concentrations. *FEMS Microbiol Lett* **2**: 65–69.
46. Zaritsky A, Woldringh CL. (1978) Chromosome replication rate and cell shape in *Escherichia coli*: lack of coupling. *J Bacteriol* **135**: 581–587.
47. Zaritsky A, Woldringh CL, Fishov I, Vischer NOE, Einav M. (1999) Varying division planes of secondary constrictions in spheroidal *Escherichia coli* cells. *Microbiology* **145**: 1015–1022.
48. Zaritsky A. (1971) Studies on DNA Replication and Cell Division in Bacteria. PhD Thesis. The University of Leicester (UK), 97 pp.
49. Meacock PA, Pritchard RH, Roberts EM. (1978) Effect of thymine concentration on cell shape in Thy- *Escherichia coli* B/r. *J Bacteriol* **133**: 320–328.

50. Zaritsky A. (1975b) On dimensional determination of rod-shaped bacteria. *J Theoret Biol* **54**: 243–248.
51. Zaritsky A, Wang P, Vischer NOE. (2011) Instructive simulation of the bacterial cell division cycle. *Microbiology* **157**: 1876–1885.
52. http://ariehz.weebly.com/uploads/2/9/6/1/29618953/video_s1_microbiology_review_2011.avi
53. Andersen KB *et al.* (52 coauthors). (2006) Honoring Ole Maalϕe. *Microbe* **1**: 210–211.
54. Zaritsky A. (1971) Proposal for Research Project on Control of Growth in Bacteria, 2 pp, posted on http://ariehz.weebly.com/uploads/2/9/6/1/29618953/proposel.pdf
55. Pritchard RH. (1960) Localized negative interference and its bearing on models of gene recombination. *Genet Res* **1**: 1–24.
56. Tresguerres EF, Nandadasa HG, Pritchard RH. (1975) Suppression of initiation negative DNA replication in strains of *Escherichia coli* by integration of the sex factor F. *J Bacteriol* **121**: 554–561.
57. Fantes PA, Grant WD, Pritchard RH, Sudbery PE, Wheals AE. (1975) The regulation of cell size and the control of mitosis. *J Theor Biol* **50**: 213–244.
58. Stent GS, Calendar R. (1978) *Molecular Genetics. An Introductory Narrative.* WH Freeman and Co. San Francisco, Second Edition, pp. 130–131. ISBN 0-7167-0048-4.
59. Wright S. (1994) *Molecular Politics: Developing American and British Regulatory Policy for Genetic Engineering, 1972–1982.* The University of Chicago Press: Chicago and London, p 349. ISBN 02-226-91065-2.
60. Berg P, Baltimore D, Brenner S, Roblin RO, Singer MF. (1975) Summary statement of the Asilomar Conference on recombinant DNA molecules. *Proc Natl Acad Sci USA* **72**: 1981–1984.
61. Reisz M. (2015) Robert Pritchard, 1930–2015. https://www.timeshighereducation.com/news/people/obituaries/robert-pritchard-1930-2015/2020006.article
62. http://www2.le.ac.uk/departments/genetics/jeffreys/biography

14 The Outsider

William (Willie) D. Donachie FRSE*

Bob was five years older than me and, as a Genetics student in Edinburgh, I knew about his pioneering work on recombination and the linear structure of the gene (with Pontecorvo in Glasgow). Several years later, after a post-doc in Arthur Pardee's lab in Princeton, and an uncomfortable couple of years back in the Genetics department in Edinburgh, I must finally have met Bob. I know that I must have because I gave a seminar at Bill Hayes' MRC Unit in London and he almost certainly was there: but I don't remember. (What I do remember is that Bill offered me a job! Out of the blue!)

In the MRC Unit we were able to work on almost anything we fancied without the distraction of looking for research funds, without having to teach and with the company of smart and congenial colleagues (at the time including Bill, Julian Gross, Stuart Glover, John Scaife and Marilyn Monk but not Bob who had already left, and many others who arrived over three happy years in London and six more in Edinburgh). Since Julian and Marilyn had bagged DNA replication, Stuart restriction and modification and John RNA synthesis, that left cell division for me! The first (and only) obvious thing about cell division in bacteria at that date (1965) was that it seemed to require prior chromosome replication and so I began what turned out to be my life's-work by looking at recovery of division in

*Professor Emeritus, University of Edinburgh (uilleam4@gmail.com).

Escherichia coli after a period of inhibition of DNA synthesis. This was immediately fruitful because it led first to explanations of "thymineless death" (induction of a cryptic prophage in some strains plus the fact that thymine starved filamentous cells of prophage-free strains are killed by plating on thymine-supplemented plates).[1] Colony counting was then the usual way of counting cells, and the only way in the Unit at the time. There was a single old microscope, in a cupboard, but no-one used it! The second thing we found was that the long featureless filamentous cells that grew in the absence of DNA synthesis had precisely located "potential" division sites that were used as soon as the DNA/mass ratio of the cells had returned to normal.[2,3] So the importance of the ratio between DNA and mass was almost immediately obvious, at least for cell division. At this time Cooper & Helmstetter were publishing their seminal work on the periods of DNA replication relative to the time of cell division in synchronous *E. coli* cells growing at a variety of different growth rates in different media at 37°C. (Getting to Bob soon!) I think that Julian presented these papers at the weekly journal club in the Unit. I was fascinated by this work and its brilliant summary as the "BCD" model of the cell cycle. Within a range of growth rates *C* (the time between initiation and termination of a round of chromosome replication: "ror") and *D* (the interval between termination of each ror and completion of cell division) were constants, but *B* (the period of growth required before initiation of each ror) was a variable dependent on growth rate. Bizarrely (as it seemed at first sight) initiation and completion of rors overlapped progressively as the generation time (*T*) became less than the interval [*C* + *D*] and a progressively increasing gap (a sort of G1) appeared between rors as *T* became greater than [*C* + *D*]. My immediate question was, "What is the common factor at the time of initiation of ror in cells growing at different rates?" Obviously many others must have asked the same question at the same time. But I appear to have got the answer first. I knew that Maaløe and his colleagues had shown that initiation of DNA replication required prior protein synthesis and "protein synthesis" in *E. coli* is almost synonymous with "increase in cell mass". I also knew that the Copenhagen group had measured average cell mass in exponential cultures of *Salmonella typhimurium* at 37°C in different media at different growth rates. My bacteriologist colleagues told me that *E. coli* and *S. typhimurium* were practically identical organisms and so I decided to calculate the cell size at initiation for (*C* + *D*) values from *E. coli* and the data on cell size and growth rate from *S. typhimurium*. Wondrously, the values for these cell masses turned out all to be the same (*Mi*) at *T* more than (*C* + *D*) minutes, or 2*Mi* or 4*Mi* in cells growing in richer media. Almost there! The last piece of the puzzle slotted into place when Cooper's & Helmstetter's model showed that the number of copies of the chromosome origin of replication (*oriC*) per cell increased with growth rate in exactly the same way. So, in an *E. coli* cell growing at 37°C in different growth media, with generation times between about 70 minutes and 20 minutes, chromosome replication is initiated at every copy of *oriC* (i.e. from 1 to 4 per cell) whenever the ratio of cell mass to number of *oriC* copies reaches a fixed value, the "initiation mass" *Mi*. Eureka!

This calculation quite accurately predicts what cell sizes will be in exponential populations of *E. coli* growing at the same temperature but different rates in various growth media, and exactly what will be the kinetics of change in cell size, DNA/mass, DNA/cell, number of copies of *oriC* per chromosome and per cell, and cell division rates during shifts from one growth medium to another or after a period of growth without DNA synthesis, or how much DNA would accumulate in cells after inhibition of cell growth.

I had been invited to give a presentation at a joint SGM/Biochem Soc symposium in Leicester in April 1967 and had submitted an abstract about something else (enzyme synthesis during the cell cycle probably). However, I was so excited about this wonderful discovery that I broke one of Pardee's wise rules: "Don't talk about your work until that work is actually in press!" After this talk, which was well received, I met Bob and his group who told me that they had been working also on the problem of simultaneous initiation of DNA replication at *oriC* during the cell cycle, and that they had formulated a model. Their model hypothesised an inhibitor of initiation that was formed in a brief pulse after initiation of each ror and then diluted by cell growth until exactly one generation later it had been sufficiently diluted to allow the next ror to begin. They had not, however, quantitatively linked this to cell size per se but could now see how to do it. I hope I have that right: in their subsequent paper[4] they thanked me for showing them how to calculate cell sizes at different growth rates. After the usual delays for refereeing etc., my own paper was published in 1968.[5]

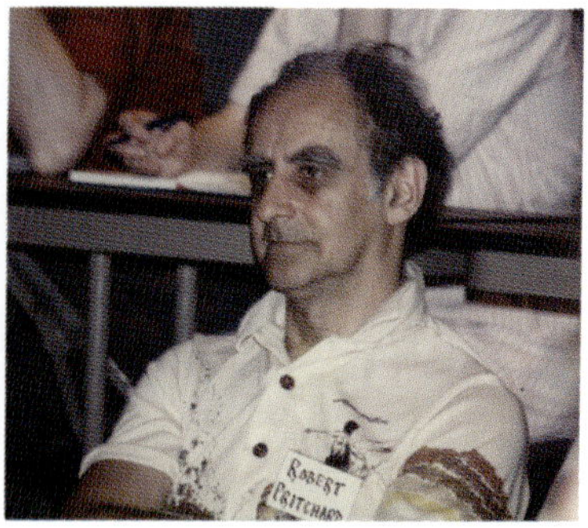

At the Ole Maaløe's 70th birthday Festschrift, 1984, Tuscaloosa, Alabama, USA.

Above: With David Freifelder.

Below: With Peter Kuempel (L) and Willie Donachie (middle).

(*Courtesy: Millie Masters-Donachie.*).

Pardee once told me: "Willie! Ideas are cheap!" and that what mattered was to do experiments. Isaac Newton also said "*Hypotheses non fingo*". However, not being either of those great scientists I did, at the end of my career collaborate with Garry Blakely to see if some sort of plausible molecular model for initiation could be cobbled together from the mass of data on the mechanism of chromosome replication that had by then accumulated.[6] We were quite pleased with this effort until after I had presented it in Uppsala, when Kurt Nordström pointed out that our model failed to explain why initiation should take place at a specific *Mi* (or initiation volume *Vi* as I preferred by then). What did please me about this model (which may still be partly plausible) was that it proposed that the time of initiation was controlled by competition between a positive initiator (Dna-ATP) and a competitive inhibitor (DnaA-ADP) for binding to multiple sites at *oriC*. Thus neither the idea of titration of *oriC* by an initiator nor the Leicester model of control by an inhibitor would be correct as originally stated but both would be required parts of this model. This is most often the outcome of arguments over hypotheses — both right and both wrong!

What is really sad is that by the time we had formulated this compound model Bob was no longer able to appreciate it. It would have been great to have argued about it over a couple of whiskies.

References

1. Donachie WD, Hobbs DG. (1967) Recovery from 'thymineless death' in *Escherichia coli* 15T-. *Biochem Biophys Res Commun* **29**: 172–177.
2. Donachie WD, Masters M, Hobbs DG. (1968) Chromosome replication and cell division in *Escherichia coli* 15T- after growth in the absence of DNA synthesis. *Nature* **219**: 1079–1080.
3. Donachie WD, Begg KJ. (1970) Growth of the bacterial cell. *Nature* **227**: 1220–1224.
4. Pritchard RH, Barth PT, Collins J. (1969) Control of DNA synthesis in bacteria. *Symp Soc Gen Microbiol* **19**: 263–297.
5. Donachie WD. (1968) Relationship between cell size and time of initiation of DNA replication. *Nature* **219**: 1077–1079.
6. Donachie WD, Blakely GW. (2003) Coupling the initiation of DNA replication to cell size in *Escherichia coli*. *Curr Op Microbiol* **6**: 146–150.

Post-script

Having read the contributions by Arieh and Peter Barth I am really glad that I published promptly! "*Ars longa, Vita brevis, Judicium fallum, Memoria labilis*": Isaac Newton.

15 Appendix: Interactions with Outsiders and Insiders — Justice to Priority Truth

Arieh Zaritsky*

My major, lifelong scientific interest, Bacterial Physiology and The Cell Cycle, was inseminated by **Bob** in my Leicester formative experience. **Peter Barth** had just left three months before my arrival hence the gossip information about previous events in the field was delivered to me by **John Collins**. These were the years at which the field was consolidated by The Copenhagen School, as mentioned in this compendium.[1,2] Particularly, cell size and macromolecular composition as a function of doubling time τ and during nutritional shift-up transitions was established by **Maalϕe**'s team in 1958[3,4] and the values of *C* and *D* periods were found to be constants over a wide range of τ's by **Helmstetter** *et al.* a decade later.[5] Together, these seminal observations led to the conclusion that the main regulatory circuit of the cell cycle, initiation of chromosome replication occurs when cell mass (or volume) reaches a value *Mi* that is directly related to the number of *oriC*'s, as was so elegantly shown by **William (Willie) Donachie**.[6] During the following decades, I've been convinced that this relationship (constant *Mi*/*oriC* copy-number) was discovered simultaneously, slightly preceded by Bob's team.[7,8] This wrong belief has slowly been shaken by cyberspace

* Ben-Gurion University of the Negev, Be'er-Sheva, Israel (ariehzar@gmail.com).

discussions between John and Peter, as well as mine with Helmstetter. These, and my own thoughts, were instrumental to grasp (yet again) that "ideas are cheap" (Willie citing **Arthur Pardee**; but see below) hence my recent references to this cornerstone of the field included all three articles.[5–7] It becomes obvious now that this phenomenon, which underlies the molecular mechanism regulating replication-initiation and hence by default the whole cell cycle, was implied in Helmstetter *et al.*[5] but explicitly and simultaneously stated and demonstrated by Willie![6] What Bob and his team of bright, young students advanced in their crucial article[7] is the idea that the mode of regulation is negative, as is common in biological systems — particularly widespread during the decade of **Jacob** & **Monod's** *lac* Operon Model.[9] As Willie articulated[,10] different models formulated to explain a phenomenon eventually converge[11]; Bob would have agreed with this view, but would have also found another detail to argue about, as he used to do during his life: refining existing hypotheses and constructing new ones, with an admirable "innovative power". In this sense, Willie and Bob could have been considered as 'academic twins'.

The "ideas are cheap" notion is true in terms of physical resources no doubt, but may be misleading intellectually because most ideas turn out to quickly be refuted in real-life hence "long-sustaining ideas are scarce" is a more appropriate concept. Lucky us, who continue to operate in the realm of the same BCD model of the Bacterial Cell Division Cycle for half a century continuously, refining[12] and exploiting it for furthering understanding of the bacterial cell.[13]

My fond memories of Willie can be traced to our numerous encounters in international conferences and joint annual unofficial meetings of Leicester/Edinburgh Geneticists during my studies. Most of all was him being my *viva* external examiner in the spring of 1971. Willie was highly sympathetic, and being familiar with the Thesis material, he was mainly interested in the project that I proposed for my next EMBO long-term fellowship[14]; he had been assigned by EMBO to referee/review that project proposal, and I shall always be grateful for kindly treating me.

This appendix cannot be complete without mentioning additional insights looking at the long-term, continued competition between various laboratories during the period under description in this series of articles. The urge to discover new phenomena or mechanisms, at least refine such, is one of the forces that drive scientists hence the frequent arguments about so-called "priority claims". Again, I find Art Pardee's (cited by Willie in a recent letter to me) that ideas "are formed socially, passing from mind to mind (often unconsciously) and being constantly reinvented and perfected", a highly important observation that should be kept in our minds — and in those of current and future generations of scientists. Furthermore, "a lifetime of experience as a scientist has indeed confirmed that ideas are of their time and "in the air" long before one person sets them down and claims 'priority'". A well-known Hebrew idiom states that (Google Translator) "Envy Increases Wisdom". This has happened between Bob, Willie, Millie, Karl, Lucien and others. When envy goes to extreme however, it breeds animosity as well. Each of us should attempt to avoid that sort of feeling! To my judgment, Bob himself was proficient in doing so.

References

1. Barth PT. (2016) This compendium, pp. 15–24.
2. Zaritsky A. (2016) This compendium, pp. 61–78.
3. Schaechter M, Maalφe O, Kjeldgaard NO. (1958) Dependency on medium and temperature of cell size and chemical composition during balanced growth of *Salmonella typhimurium*. *J Gen Microbiol* **19**: 592–606.
4. Kjeldgaard NO, Schaechter M, Maalφe O. (1958). The transition between different physiological states during balanced growth of *Salmonella typhimurium*. *J Gen Microbiol* **19**: 607–616.
5. Helmstetter CE, Cooper S, Pierucci O, Revelas E. (1968) On the bacterial life sequence. *Cold Spring Harbor Symp Quant Biol* **33**: 809–822.
6. Donachie WD. (1968) Relationships between cell size and time of initiation of DNA replication. *Nature* **219**: 1077–1079.
7. Pritchard RH, Barth PT, Collins J. (1969) Control of DNA synthesis in bacteria. *Microbial Growth. Symp Soc Gen Microbiol* **19**: 263–297.
8. Pritchard RH. (1968) Control of DNA synthesis in bacteria. *Heredity* **23**: 472–473. (Abstract).
9. Jacob F, Monod J. (1961) Genetic regulatory mechanisms in the synthesis of proteins. *J Mol Biol* **3**: 318–356.
10. Donachie WD. (2016) This compendium, pp. 79–83.
11. Donachie WD, Blakely GW. (2003) Coupling the initiation of DNA replication to cell size in *Escherichia coli*. *Curr Op Microbiol* **6**: 146–150.
12. Amir A. (2014) Cell size regulation in bacteria. *Phys Rev Lett* **112**: 208102.

13. Zaritsky A. (2015) Cell shape homeostasis in *Escherichia coli* is driven by growth, division and nucleoid complexity. *Biophys J* **109**: 63–80 (+ 2pp Supporting Material — L01, L02).
14. Zaritsky A. (1971) http://ariehz.weebly.com/uploads/2/9/6/1/29618953/proposel.pdf

16 Bacterial and Plasmid Replication: Some Memories of the Early Days

Millicent Masters*

In 1965 I joined Bill Hayes' Bacterial Genetics Unit at Hammersmith Hospital supported by an NIH Postdoctoral Fellowship. My PhD work at UC Berkeley and Princeton was concerned with gene expression during the *Bacillus subtilis* cell cycle. At Princeton we interacted with the Sueoka group in which Hiroshi Yoshikawa was engaged in the elegant demonstration that *B. subtilis* had a fixed origin of chromosome replication. This was shown by calculating recombination frequency ratios for a variety of markers, after transformation with DNA extracted from cultures growing in supplemented minimal medium (g = approx. 30 min) or stationary phase cultures. The result was a 2:1 gradient in transformant frequency ratio. This approach was based on the hypothesis, supported by their results, that replication was sequential from a fixed start point or points and proceeded at a fixed rate in the growing cultures, while all chromosomes were fully replicated in the stationary ones (*Proc Natl Acad Sci USA* **49**: 559–566, 1963.)

When I arrived in London I found a unit mostly concerned with *Escherichia coli* and thought that it would be great to somehow extend the *subtilis* genetic methodology to *coli*. Luckily, during part of my three

*Edinburgh University (m.masters@ed.ac.uk).

years in London, Lucien Caro and Claire Berg were on sabbatical at Hammersmith and were using P1 transduction to measure gene frequency. They were kind enough to share the mysteries of P1 transduction methodology with me and I conceived the notion of applying the Yoshikawa and Sueoka method to *E. coli* using P1 as the vector to carry DNA from cultures growing at different rates to recipient cells. This work took several years to complete, by which time I had relocated with the MRC Unit to Edinburgh (Masters M, Broda P. (1971) Evidence for the bidirectional replication of the *E. coli* chromosome, *Nat. New Biol.* **232**: 137–140). Bidirectionality was clearly demonstrated and confirmed by us biochemically the following year (McKenna WG, Masters M. (1972) *Nature New Biol* **240**: 536–539). Further confirmation and more accurate origin positioning were provided by the Caro lab (Bird R, Louarn J, Martuscelli J, Caro L. (1972) *J Molec Biol* **70**: 549–566) amongst others. It would have been sometime during the late 60s or early 70s that I would have met Bob. I do not remember exactly when, but it was at a time when directionality of bacterial chromosome replication was a much aired question. Indeed, at a conference in Oak Ridge a group of us got together and reached an entirely incorrect consensus conclusion (I think we had replication starting at about 60 min and proceeding unidirectionally counter clockwise!).

I mostly recall Bob from the many plasmid meetings that we both attended where he was speaker of choice on the negative control of replication in both *coli* and plasmids. Timing of plasmid replication in the cell cycle was also of interest to many. Most of the early publications used synchronously dividing cultures, and measured β-galactosidase inducibility, which would be expected to double after plasmid duplication. They mostly concluded that inducibility and hence plasmid replication occurred at a particular time in the cell cycle, although there was no agreement as to when that was. In 1975 Bob and his group (Pritchard R.H., Chandler M.C., Collins J. (1975) Independence of F replication and chromosome replication in *E. coli. Mol Gen Genet* **138**: 143–155), working with exponentially growing F′*lac* cultures, measured β-gal inducibility at different chromosome replication velocities and growth rates and concluded that F′ replication was not coupled to any stage of the DNA replication or cell division cycle. My small foray into timing of plasmid replication came with my second PhD student Vala Andresdottir, from Iceland. As one who had worked with synchronized cultures during my own doctoral studies I

The photo of "The young, bearded, striking Bob" must be included!
(Cited from Millie Masters.)

was deeply suspicious of the use of synchronous cultures to draw conclusions about the normal timing of cell cycle events, since it was difficult to exclude the possibility that the synchronizing procedure itself had caused phasing absent from unperturbed exponentially growing cultures. This led to us purchasing a zonal rotor which we were able to use to separate exponential steady state cultures, following a brief period of IPTG induction, into aliquots with the same cell size and hence age. Our work, begun well before Bob's was published, also came to the conclusion that F'*lac* replication was not phased, but occurred at random times during the cell cycle (Andresdottir V. Masters M. (1978)). Evidence that F'*lac* replicates asynchronously during the cell cycle of *Escherichia coli* B/r. *Mol Gen Genet* **163**: 205–212).

On a more personal note, it was always nice, over those years, to see Bob's familiar face as I arrived at a conference at some far-flung place. Oh, and a memory of an unanticipated skill, Bob was most accomplished at waltzing!

17 My Pritchard Years: From Replication to Pinball and Back Again

Michael (Mick) Chandler*

I arrived at the Leicester Genetics Department in 1970 with a degree in Biochemistry from the University of Sussex following a graduate year at the University of California, Santa Cruz, peppered by anti-Vietnam war demonstrations, student demonstrations and rock concerts. I had no idea what to expect.

Bob Pritchard was recommended to me as a good PhD advisor and as one of the best microbial geneticists around by Neville Symonds (University of Sussex, d. 2014). Astonishingly, when I later wrote to Bob, he accepted me as a graduate student without an interview (telephone calls between England and California were expensive in the early 1970s!) and simply on the recommendation of Neville and my mentors at Santa Cruz, Rick Davern (d. 1988), Eric Terzaghi and Harry Noller — probably a dangerous thing to do.

On meeting Bob for the first time, I remember thinking how ferocious he looked with strong eyebrows and a very penetrating gaze. He simply could not accept intellectual sloppiness and with this belief, he provided a very exciting rigorous scientific environment. I very much enjoyed this feisty side of his character both in his science and in his politics. While I

*IScentral: 45 Ave. de l'Aeropostale, 31520, Ramonville, France (mike.chandler@ibcg.biotoul.fr, http://www-is.biotoul.fr).

Leicester railway station (Photograph courtesy of P.A. Meacock).

was a graduate student, Bob became more and more involved in local (and national) politics and, with his inimitable logic and tightly argued views, was drawn to the Liberal party (before it became the Liberal Democrats). His public-spiritedness led him into a variety of crusades. One of these, which I remember very well, was to save the impressive high Victorian Leicester railway station (picture included) from demolition, a cause which he finally won with great pride.

In his liberal (not necessary Liberal) way of approaching science, Bob threw me in at the deep end, pushing me towards various questions concerning gene expression, DNA replication and bacterial plasmids — and finally my degree turned out be a cocktail of all three. I ended up investigating the effect of relative gene concentration on the contribution of its product to overall protein composition of the cell. This used the trick developed in the Pritchard lab of changing the time taken to replicate the *E. coli* chromosome by changing the thymine concentration in the medium of a thymine low-requiring strain. The results, perhaps unsurprisingly, indicated that for unregulated genes, this was proportional to the copy number. Since copy number depends on the physical position of the gene on the chromosome, this demonstrated that genes located closer to the origin of replication contributed more than those at the replication terminus. The results implied that *E. coli* chromosome replication was bidirectional, although I remember having many hours of heated discussion with Bob on whether we had shown that chromosome replication was bidirectional or whether we had shown that gene expression is proportional to gene dosage. I thought the former was more exciting, Bob, the latter. This was 1972–73, the direction of replication had not yet been demonstrated and I had gone through the Cairn's experiments and had decided that the results were more consistent with bidirectional than unidirectional replication. We then exploited this approach of gene dosage measurement by gene expression to determine that replication of the plasmid F was not coupled to the *E.coli* replication cycle.

I am sure that he believed that my year in California had provided me with much more technical baggage to undertake experimental microbial genetics than I actually possessed. This was one of Bob's strengths, to allow juniors the freedom to follow their own noses. My major recollections during the three years of my degree (1970–73) were of passionate and often heated but, in the end, good humored scientific discussions with Bob; of being pinned to my bench in animated arguments; of undertaking highly choreographed multi-step experiments in which the steps often slipped out of phase resulting in canisters of glass pipettes flying through the air; of brewing beer; of animated parties, and especially being taught by Bob how to open a bottle of wine without a corkscrew. It was an excellent trick... but not to be recommended unless you have a good plasterer at hand!

This was quite early in the life of the Leicester Genetics Department which Bob, as one of the youngest Professors in the country, was building and the nascent Department had begun to gather a reputation. It was also a time when many people passed through Leicester and this gave a real element of excitement for a young graduate student and gave me the opportunity to meet scientists such as Willie Donachie, Millicent Masters and Ezra Yagil, who have become friends and with whom I still maintain contact, and of course Kurt Nordstrom.

I will always remember Bob's kindness and, as one example, I include one incident which occurred during my viva when the external examiner, Robin Rowbury (d. 2012), asked me what I thought of *dnaE* (a question on which I faltered badly) and to which Bob answered simply "We don't"!

It was sad to move on but, after obtaining my degree and stimulated by Bob's keen interest in "Integrative Suppression" (or "suppressive integration" as he liked to call it because this was more grammatically correct in English), a phenomenon in which bacterial chromosome replication was thought to be driven by an integrated plasmid, I left for a Royal Society postdoctoral fellowship with Lucien Caro (d. 2015) in the Department of Molecular Biology at the University of Geneva, where more direct techniques of DNA-DNA hybridization had been developed to analyze this type of problem. Almost the first set of experiments I undertook in Geneva was to use DNA-DNA hybridization to measure the replication time of the *E. coli* chromosome as a function of growth rate and then to determine that replication of the plasmid-like phage, P1 prophage, like that of F, was not coupled to the cell cycle. These early analyses were directly inspired by Bob. The experiments were a prelude to our demonstration (again directly inspired by Bob) that chromosome replication, both in *E. coli* cultures synchronized for their replication and in exponentially grown cultures, can be driven by an integrated plasmid. This led directly to my subsequent interests in plasmids and their most significant components: the transposable elements.

During this long period (it turned out to be a nine year postdoc!), we were lucky to have Bob visit us with his son Simon for an extended period and, apart from re-establishing animated scientific discussions and enjoying the surrounding Geneva countryside, I feel that my one major contribution to Bob's life was initiating his addiction to the game of pinball in the local bistro, Le Grand Marché, in the mid- to late-70s!

I treasure my last memory of him before he fell ill at the beginning of the "naughties". Sitting in the living room of his large house he took great delight in recounting an incident in which he put a provocative and entirely unreasonable young journalist from the Leicester Mercury in his place. Briefly, the story involved an accusation that Bob supported the use of various illegal substances (a "liberal" and purely logical rather than a "Liberal" stand). In a telephone conversation, Bob managed to obtain an admission from the journalist (who had called him Prof. Bob "spliff" Pritchard) to having indeed consumed such substances in the past. With a great glint in his eye, Bob explained to me "I told him that I had recorded the entire conversation and that if he did not withdraw his story, I would

At the party on behalf of Pontecorvo's 80th birthday, 15 December, 1987.
Left-to-right: Obaid Siddiqi, Bob Pritchard, Ted Forbes, Jennifer Dee.

send it to his editor-in-chief". Of course Bob hadn't and didn't but the threat did the trick.

For me and for many others, it was a very great privilege to have known and worked with Bob and, via his mentoring, to have learnt to think.

18 My Time in Bob's Lab (1975–78)

Ramón Díaz Orejas*

Choice of Lab and Arrival to Leicester

I first met Bob in Madrid, one year before arriving to Leicester, during a visit he did to the laboratory of José Luis Cánovas at the CIB (CSIC), where I was working for my PhD under the direct supervision of G. de Torrontegui. José Luis was a former postdoctoral student in Hans Kornberg's lab in Leicester and then in Roger Stanier's group in Berkeley; after returning to Madrid he was moving from studies on bacterial metabolism and its regulation to cell cycle studies. Due to this, Maria Elena Fernandez Tresguerres, his first PhD student, had moved to Bob's laboratory, where she made an important contribution on chromosomal replication driven by an integrated F plasmid (integrative suppression). Soon after returning from Leicester she was appointed as staff scientist at the CIB and started working in José Luis's group on bacterial cell division. This incorporation consolidated the thematic change in this group and, jointly with the visit of Bob to Madrid, determined my post-doctoral in the same laboratory.

I arrived to Leicester from Madrid with my wife Pilar early in January 1975. Barry Holland was helping us to find accommodation, first in a close-by hotel and then in the "International Student Hostel", 30th St. James Road.

*Centro de Investigaciones Biológicas (CIB, CSIC) (diazorejasr@gmail.com).

There we lived happily during the next four years!. At that time I hardly could understand the BBC speakers, thus communication was difficult and the strong Midland accent was not helping! However, my English soon improved. I remember that Peter Meacock was finishing his PhD in Bob's lab, and then Grantley Lycett, the last PhD student of Bob, arrived. Both of them were very helpful and friendly and helped me to overcome this important barrier. I remember that Bob took me and Pilar for lunch at the Senior Common Room; I liked the kind of English-club atmosphere created by people reading the newspapers after lunch. We kept going there for lunch and one of the times we met, while waiting for food, Joaquin Dopazo, a Spanish student finishing his PhD in economy; soon after, we were in contact with many nationals in town. The Hostel at St. James Rd. was also a source of stimulating contacts. Many students from different countries, some of them from Spanish-speaking countries, were hosted there. Tony Nicolaidis, then finishing his PhD in Barry Holland's lab, was our neighbor. We shared the kitchen with him and with a charming Turkish girl. Cooking together was an occasion to talk and to exchange recipes! His fluent English (and conversation!) helped us to improve our poor English. I treasure good memories of the atmosphere that so many different students created there; some of them stayed as good friends during all these years.

Cloning *oriC*: A P1 Transduction Approach

Bob's ideas on control of chromosomal replication regulated by a negative effector changed the paradigm established by the replicon model of Jacob *et al.* Bob's model, the "Inhibitor Dilution Model", proposed that chromosomal replication was regulated at the initiation stage by a repressor that was synthesized in a pulse during each round of replication; this repressor was diluted during subsequent cell growth and a new round of replication occurred periodically when a defined initiation mass was reached. It was thought that cloning *oriC* would open an important experimental window to search for the replication inhibitor and its site of action, to analyse other factors involved in control of replication and to evaluate, in a new context, the coupling between replication and other cell cycle events. This was an objective in Bob's lab as well as in other laboratories interested in cell cycle regulation.

As a strategy to clone *oriC* we first resorted to P1 mediated transduction: we were using as donor a P1 lysate propagated in a wild type *E. coli* strain, and as recipients, recombination deficient strains carrying mutations close to *oriC* that prevented the utilization of particular carbon sources; transductants that complemented these mutations were selected, with the hope that if *oriC* was co-transduced with the appropriated nutritional marker, this transductant could be established as an autonomous *oriC* replicon. This minimalist and elegant approach did not work in my hands. I remember the interest of Bob Iyer, then a visiting scientist in Barry's lab, to follow up this project. He was arriving early in the morning to our laboratory to follow up this project and to analyse the results with me. I am very grateful to many people that were very helpful in different moments of my time in Leicester, of course Peter Meacock and Grantley Lycett with whom I was sharing the lab, but I am remembering this second Bob also, with veneration and deep gratitude. In addition to being an excellent and prestigious scientist, he was for me, also some kind of a paternal figure.

Plan B: The Genetic Engineering Approach

Then P. Meacock finished his PhD work and moved to Stanley Cohen's lab in USA, to use the newly developed genetic engineering technologies in an important biotechnological project. Stanley Cohen's laboratory pioneered important advances in these technologies; there the first transformation protocol for *E. coli* was developed and the first cloning vehicle, pSC101, was isolated. There also, a fragment containing the replication origin of F plasmid, a mini-*oriF* plasmid, was isolated as an autonomous replicon conferring resistance to ampicillin by K. Timmis and F. Cabello. The approach used was to fragment the whole F plasmid with restriction enzymes and to bind the fragments to a fragment containing a β-lactamase gene (resistance to ampicillin, Ap fragment) but unable to replicate. This encouraged Bob to propose a plan B: to scale up the process to isolate a fragment of the *E. coli* chromosome containing the origin of replication, *oriC*. Key to our new project was the availability of pSC122, a plasmid recombinant constructed in S. Cohen's lab that included DNA of two far away strains: *Salmonella* (pSC101 vector) and *S. aureus* (Ap fragment).

The new genetic technologies were arriving at a few Universities in UK, and books of protocols and enzymes required were not readily available from commercial sources. We mentioned in the acknowledgments of our first publication that *EcoR1* used in our constructions was a gift from K. Murray in Edinburgh University and I think ligase, used later on in our experiments, was also a present. Seen in perspective it was a great responsibility and certainly a great honour that Bob trusted me to introduce these important technologies in the Department; particularly after the failure of the first approach or perhaps due to this!

Bob proposed to start the project cloning in a multicopy ColE1 vector the *EcoR1* Ap fragment of pSC122 (pSC101-Ap). As the copy number ColE1 is higher than the one of pSC101, a ColE1-Ap recombinant was a more efficient source of the Ap fragment than pSC122; in addition ColE1 could be further amplified in the presence of chloramphenicol. ColE1 was thus amplified, then this DNA was isolated using CsCl gradient and it was cleaved at its unique *EcoR1* site. In our experiment pSC122 was digested also with *EcoR1* to separate the vector and the Ap fragment and finally these fragments were mixed with an excess of linearized ColE1 and used to transform *E. coli* K12 C600 strain. Selection for ampicillin-resistant clones gave several transformants, one of which has the signature of a ColE1-Ap recombinant: it was immune to colicin E1 but unable to produce this colicin; consistently digestion with *EcoR1* gave fragments of sizes corresponding to the ColE1 vector and the Ap fragment. We called this construction LG1 (Leicester Genetic 1), the first recombinant made in Leicester University; interestingly, it was obtained without using ligase to join *in vitro* the *EcoR1* fragments. Instead, these fragments were joined *in vivo* by cell lygase following internalization of the DNA. This economic and minimalistic approach was quite inefficient and of course become obsolete when T4 lygase was readily available.

The ColE1-Ap recombinant was used as an enriched source of the Ap fragment. We prepared ^{3}H labelled pLG1, cleaved it with *EcoR1* and the ColE1 and Ap fragments were further separated in CsCl gradients by their different buoyant densities. The Ap containing fractions were pooled, dialyzed and subsequently used as a hook to isolate a chromosomal fragment that could drive autonomous replication of Ap (*ori-Ap* replicon).

The Isolation of the First Origin Fragment: *oriJ*

To this aim, DNA of C600 was isolated by Marmur's method and then it was digested with *EcoR1*; the collection of chromosomal fragments originated were ligated to the Ap fragment. This time ligation was optimized, treating the DNA mixture with T4 lygase. The ligation mixture was transformed in the recombination deficient *E. coli* strain AB2463 and transformants were selected in minimal medium containing ampicillin. In this way three ampicillin resistant transformants, hopefully containing an *oriC EcoR1* fragment, were isolated. The size of the probable *EcoR1* fragment containing *oriC* was already proposed to be 9.7 kb in base to the restriction map of F-prime derived plasmids from KL228, an Hfr strain transferring early the *oriC* region. To test if a fragment of this size was present in our recombinants, we isolated the plasmids containing the three ampicillin resistant clones and digested them with EcoR1; we found that the three of them were identical and that they originated two fragments of 7.4 kb, the size of the Ap fragment, and of 13.6 kb, not of 9.7 kb, the size expected for *oriC*!. These differences puzzled us but still it was possible that the 13.6 kb fragment could contain *oriC*: as we wrote in the article describing our results "it was conceivable that different K12 strains have different restriction target maps in the neighbourhood of the chromosomal origin". This possibility was evaluated by Southern blotting using labelled *ori-Ap* plasmid to identify a homologous *EcoR1* fragment of KL228 chromosome, the strain that originated F-prime containing the 9.7 kb *oriC* fragment. We could not detect a fragment of this size. Instead a fragment of 13.6 kb was detected, thus ruling out that this fragment could contain *oriJ-Ap*. Isolation of these clones happened in the Silver Jubilee week. Due to this and to Brian Spratt's suggestion I called the construction *oriC-Ap*. Alec Jeffreys helped me to do the first Southern blotting experiments that were so important in our work, and to derive the first chromosomal maps.

The isolation of *oriJ* suggested that we had found a different fragment of the *E. coli* chromosome having the potential to act as an origin of replication. This result was exciting in itself because it suggested the existence of more than a functional replication origin in the *E. coli* chromosome. However, it left us with two problems: to identify the source of the mysterious *oriJ* origin and to isolate the *oriC-Ap* recombinant to which we first aimed!

Identfication of *oriJ*

To cut the story short, the so called *oriJ* origin was identified as the origin of replication of a defective prophage called Rac that was proposed by B. Low to be close to the terminus of chromosomal replication; ironically *oriJ* was located as far as possible from *oriC*! The clue to this identification came from the indication by D. Burt, at that time working in the laboratory of B. Brammar, of the existence of Rac, a deficient prophage that is present, close to the replication terminus, in most *E. coli* strains. This included C600 that we were using as a source of *E. coli* chromosomal DNA but not AB2463, the *recA* strain that we were using as recipient to rescue possible *ori-Ap* recombinants! As the replication repressor of Rac was absent in AB2463, this strain could allow autonomous replication of possible recombinants containing the origin of replication of Rac. This turned out to be the case: we found that *oriJ-Ap* hybridized with F-prime 123 that contained DNA of this region but not with F DNA! λ DNA and the DNAs of P1 and P22 phages were discarded as possible source of *oriJ* as they did not cross-hybridize with *oriJ-Ap*. Further support for this identification came from transformation experiments: we reasoned that if *oriJ* was the origin of replication of Rac, transformation of *oriJ-Ap* recombinants should be prevented or greatly reduced in C600 due to expression of the Rac repressor. Indeed, very few colonies appeared in this experiment; after curing the plasmid in these rare transformants, we found by Southern blotting that they contained an extensive deletion of sequences homologous to *oriJ* and that they were efficiently transformed by *oriJ-Ap* plasmid. This and further work made in collaboration with K. Kaiser, at the time working for his PhD on defective lambdoid phages in *E. coli* K12 in N. Murray's lab, established clearly *oriJ* as the origin of replication of Rac prophage. We suspected that the preferential rescue of *oriJ-Ap* recombinants in our experiment (three identical clones were isolated and none of them contained *oriC*!) could have been favoured by induction of the prophage in some of the C600 cells, thus enriching the DNA preparations in Rac (*oriJ*) sequences.

Isolation of Truly *oriC* Recombinants

Characterization of *oriJ-Ap* recombinants gave us a clue to isolate *oriC* plasmid: we noted that in rich medium there was selection against colonies containing *oriJ-Ap*. Therefore we repeated the experiment selecting transformants in rich LB-agar medium containing ampicillin. In the experiment we were using again C600 as a source of chromosomal DNA, but following Bob's recommendation, instead of AB2463 I used as recipient LC248, a Rac$^+$ strain with an integrated F replicon. The integrated F replicon would provide a way-out in case the extra sequences provided by an *oriC-Ap* plasmid could compete replication from *oriC* in the chromosome and the Rac prophage was counter-selecting *oriJ* recombinants due to the Rac immunity determinant. Later on it was found that this precaution was not required. This indicated that *oriC* recombinants did not prevent chromosomal replication. In addition, and different to *oriJ* recombinants, the possible immunity associated with a replication repressor was not preventing an efficient establishment of *oriC* recombinants. This time five identical *oriC-Ap* recombinants were isolated. Restriction analysis indicated that all of them contained two *EcoR1* fragments with sizes corresponding to *oriC* and Ap (9.7 and 7.4 kb respectively). Identity of the *oriC* fragment was further confirmed by Southern's hybridization: *oriC-Ap* plasmid hybridized with a chromosomal *EcoR1* fragment of 9.7 kb and with a fragment of the same size carried by a λ-*oriC* recombinant that was kindly sent to us by K. von Meyenburg. By then, *oriC* recombinants were isolated independently, first by S. Yasuda and Y. Hirota in Japan, and shortly after by W. Messer in Berlin, K. von Meyenburg in Denmark and by us in Leicester.

The Last Chapter

Our work was published in Nature, on October 12, 1978, an important religious festivity in Spain ("El Pilar"); it was entitled "cloning of replication origins from the *E. coli* K12 chromosome".[1] Soon after the first publication, we published a back to back paper with K. Kaiser and N. Murray in Molecular General Genetics detailing respectively the identification of *oriJ* as the origin of Rac prophage in the chromosome[2] and the physical characterization of this prophage.[3] I published later on, jointly with K. Kaiser, a third manuscript on the presence of a preferential attachment site for the related lambdoid phage lambda reverse (λrev) in Rac$^-$ strains.[4]

The cloning of *oriC* and *oriJ* in Bob Pritchard's laboratory established the recombinant DNA technologies at the University of Leicester. The incorporation of these technologies were very much potentiated due to the appointment of Alec Jeffrey to the Department of Genetics and of Bill Brammar to the nearby Department of Biochemistry. The discovery in 1984 by Alec Jeffrey of "fingerprinting", "a powerful and elegant way of distinguishing between individuals", had an enormous social impact and made world famous the Genetics Department. Since then, it continued to lead pioneering developments in different fields of Genetics.

Due to this work and to the importance of these, then new, genetic technologies I got a permanent position at the CSIC in Spain. No matter I was feeling better than ever in the Department, and despite offers to continue there, we (Pilar and I), felt the need of a change after four intense years in Leicester. Then a quite attractive offer was opened in Kurt Nordström's group in Denmark: to develop an *in vitro* replication system for plasmid R1 that could lead to the identification and biochemical assay of the replication regulator of this plasmid and of its replication initiation protein. This project was maintaining the focus on control of replication but using a plasmid replicon, R1, in which the first mutants that allowed the genetic analysis of control of DNA replication were identified. In addition, work on *in vitro* replication of DNA had appealed to me from the first time that, as an undergraduate, I heard the fascinating story of the *in vitro* copy of fX174 DNA by DNA Polymerase in Arthur Kornberg's laboratory. Knowing the strong prestige of Bob and the close links of Kurt's and Bob's laboratories I knew that I had a good chance to get the job. This turned out to be the case, and Bob, always favouring freedom and respecting people, was not opposed to my departure. This turned out again to be a good adventure that ended with the development in Kurt Nordström's and Walter Staudenbauer's laboratories respectively of *in vitro* replication systems for plasmid R1 and for the broad-host range replicon RSF1010. The development of the *in vitro* replication system for plasmid R1 was the first report of the coupling of the three fundamental reactions of genetic information transfer (transcription, translation and replication) in a cell-free system and eventually this gave my second publication in Nature. The development of the RSF1010 *in vitro* replication system, published in NAR, gave important clues to understand the broad-host range character of this promiscuous replicon that was underlined in my first publication and then in subsequent work made at the

Department of Prof. Schuster at the Max Planck Institute for Molecular Genetics in Berlin. Back in Madrid in 1982, I continued the studies on plasmid replication with an emphasis on replication initiation proteins, and later on toxin-antitoxin (TAs) systems following our discovery in plasmid R1 of *kis-kid*, one of the first TAs found. Interestingly, the *kis-kid* system is coordinated with the basic replicon of the R1 plasmid. Certainly my experience in Bob's lab was at the base of these and of other gratifying experiences in my scientific career.

Bob's Farewell Present

I keep in my house in a selected place, a map of Leicestershire that Bob gave me as a present when I left the Genetics Department. He only said a big "THANKS" when he handed it to me. I can remember perfectly the exact place in the corridor of the Department, close to the stairs, in which he pronounced these words. Since then I keep hearing this big "THANKS"; this memory reminds me of his appreciation and it has been a powerful stimulus to move ahead whenever I was in front of a new challenge. Beyond this, his way of doing science and the way he trusted in and treated people has been an inspiration that shaped my own style as scientist and person. I hope that, wherever he is, he could hear my big "THANKS to you Bob".

Peace and love to this wise and brave man!

References

1. Díaz R and Pritchard RH. (1978) Cloning of replication origins from the E. coli K12 chromosome. *Nature* **275**: 561–564.
2. Díaz R, Barnsley P and Pritchard RH. (1979) Location and characterisation of a new replication origin in the E. coli K12 chromosome. *Mol Gen Genet* **175**: 151–157.
3. Kaiser K and Murray NE. (1979) Physical characterization of the "Rac prophage" in E. coli K12. *Mol Gen Genet* **175**: 159–174.
4. Díaz R and Kaiser K. (1981) Rac- E. coli K12 strains carry a preferential attachment site for λrev. *Mol Gen Genet* **183**: 484–489.

19 Bob Pritchard, Beach Bum (1980/1)

Douglas W. Smith*

An EMBO workshop held in September 1979 at the University of Leicester, on DNA replication and cell partitioning, was my first opportunity to get to know the charismatic and somewhat eccentric scientist Robert "Bob" Pritchard. As founding Professor and Chair of the Department of Genetics, Bob had organized a scientific meeting that drew scientists from around the world. My laboratory had taken an approach to understanding the nucleotide structure of the bacterial origin of DNA replication which focused on the proposition: if replication origins from other enteric bacteria similar to *Escherichia coli* could function as origins on plasmids in *E. coli* and if their nucleotide sequences were different from that of *E. coli*, then **conserved** nucleotides would indicate possible **required** nucleotides for function of the DNA sequence as a bacterial origin of DNA replication, at least in *E. coli*. At this meeting, I presented our results from such efforts with *Salmonella typhimurium*. Nucleotide differences were present between the two origins but they were rather few, and it was unlikely that all of the conserved nucleotides were essential. For this approach to really be informative, cloning and analysis of such origins from bacteria considerably more distant from *E. coli* than *S. typhimurium* would be needed.

*University of California, San Diego, La Jolla, CA, USA (dsmith@ucsd.edu).

Nevertheless, this approach generated considerable interest at the meeting, and Bob was particularly astute in his critical appraisal. He agreed with the above-mentioned caveat in this approach, and noted that even if reduced to a "minimal" set of "essential" nucleotides, one would have no proof that one had an actual minimal set of conserved nucleotides, nor that they were all essential for function. One did not even know that true bacterial origins of DNA replication were cloned on the plasmids; one only knew that these cloned segments functioned as such origins for initiation of DNA replication in *E. coli*.

Bob and I, and others, had a wonderful time discussing all such matters, and I was convinced that our approach was worthwhile, even with the caveats. At that time, I asked Bob if he would take a sabbatical leave with me in San Diego in the Biology Department at UCSD (University of California, San Diego), enjoying the excellent science and wonderful ambiance found there. The idea was appealing to Bob, and a year later, in the summer of 1980, Bob arrived with his wife Suzi, and young son to spend a sabbatical year.

Bob had a free rein to do whatever he wanted to in my laboratory. He had a desk, had some bench space, parking permit, keys. Hard to believe, but neither microcomputers nor email yet existed! One did "word processing" using a "smart terminal" and a Unix or VAX minicomputer in the Computer Center. Bob was not interested in any computer accounts. In fact, as it developed, he showed little interest in the bench space or in doing experiments by himself! Rather, Bob was of the breed of molecular biologists who mainly think and discuss, think about the large issues and how to tackle them experimentally, then discuss and refine details. He often participated in detailed design of experiments, letting others do the actual experiments. Bob was really good at seeing alternative explanations to experimental results and design, and he was sought out by all lab members to help develop control experiments to rule out (or in) such alternatives.

So, as the year developed, Bob would come to the lab "up on the hill" above La Jolla in San Diego essentially every day and would immerse himself in the goings-on of the lab. He liked his office, a shared facility, because he then routinely had daily interactions with lab members. He would discuss science with all people; undergraduate students were as important as senior postdoctoral fellows. He also did not hesitate to discuss other topics, such as politics and economics. He contributed extensively at our weekly group meetings, and "took his turn" at leading such meetings. We had an excellent group, largely led by Judith Zyskind who subsequently became Professor of Biology at San Diego State University. The quality of science in the lab was greatly enhanced and discussions stimulated and expanded during that year because of Bob's presence and the contributions he made.

While here on sabbatical, Bob and his family also wished to enjoy the attributes of life in sunny southern California and to partake in the cultural aspects offered by San Diego and the United States. When I asked if I could assist in buying a car or in finding a home for them, it was clear that they had some specific ideas in mind. First, they found a home right on the beach, about halfway between UCSD and downtown San Diego. Suzi, it would turn out, would spend much of her time downtown, while Bob would come to UCSD. The home they rented was a two-bedroom flat in Mission Beach, a narrow corridor suburb between the Pacific Ocean and Mission Bay, about four blocks wide, and full of beach-oriented folks. Their rented home faced directly the ocean, with only a walkway called the Boardwalk and accompanying concrete fence separating them from the beach and the ocean. This meant a 2-minute direct walk to jump into the ocean for Bob and family, and, with boogie boards and fins, this became a daily activity.

Still, the population density is high in Mission Beach, life focuses on surfing and beach activities, with much drinking and partying; hardly what one might expect for the choice of Professor and Chair of Genetics from the University of Leicester! Bob and his family nevertheless had a wonderful time, from all I could tell. It was rather amazing at how well he and his family "fitted in" with other members of this community. They usually wore shorts, T-shirts and flip-flop sandals, and one had no idea they weren't the usual sort of beach people. This is strong testament to the non-prejudice of the family, with none of the hierarchy in attitudes and behavior sometimes seen in British circles. This was a time when I was distance running, and I would sometimes run by on the boardwalk and stop for a visit with the family or whoever was there. Everyone seemed very happy, as they all pursued their lives as "beach bums".

Second, Bob found a car. My experience in Europe as a postdoctoral fellow in Germany was with a small Renault R4, a 5-door, high mileage little car that could drive through the 'alte Stadt' and the like. Perhaps this is what Bob and family had in Leicester. However, their view of American culture was a Detroit car. They wanted one of the "real" Detroit cars: a large car, a real gas-guzzler. Gas was cheap compared with England; why not enjoy the luxury of a large, comfortable car? So Bob found a Mercury "M" car, from the 1950s! This is a car that was wide, with low, huge seats, lots of chrome, and fancy non-functional detail and features. "Mercs" were popular when I was growing up; one never knew for sure what the "M" stood for: "muscle" car? "monster" car? Either definition would fit the bill. It was amusing to look out into the UCSD parking lot when Bob was at the university and see this huge car among all the other relatively small cars. Bob liked to occupy two parking spaces with this car, just in order to get out of the car. I often wondered where he parked the car near his home, since there is so little parking space in Mission Beach. But Bob and Suzi both had a great time with this car!

A word about Bob's wife Suzi. She was an extraordinary person, full of life, very active, flamboyant, a person who complemented and augmented the personality of Bob. But she was more: she was a highly talented actress. Part of the reason for choice of homes was so that she would have ready access to downtown areas where theater was to be found. She was attracted to the philosophy and productions of the San Diego company "Three's Company", a major theater company in San Diego. And she was good enough to become a member of the company! At least, she appeared in productions of the company throughout the year. And she was very good, or so she appeared to me. I wish I had had more time to get to know her better.

Bob, it turned out, had at least three real passions in life: science, liberal politics, and pinball machines. I first saw Bob's ability with pinball machines in Leicester during the 1979 EMBO Workshop when, on the last night, we went to a favorite local pub for "festivities"; they had at least one pinball machine and Bob was really good, much better than us and as good as any of the local, excellent players. And he was well-known locally for these abilities. Although Professor and Chair at the University, he was totally at ease and open-minded with these locals, and they with him. This

In La Jolla, 1980. Suzi Pritchard (*front left*), Bob Pritchard (*behind*), Caroline and Jane Barth.

was a major feature of Bob's personality: no prejudice, no one-upmanship, total open-mindedness.

This all carried over to his pinball hobby in Mission Beach. He liked going at times to one of the local bars that only had "locals", with very few tourists and out-of-towners. He became known, liked, and accepted by regulars of this bar, and liked to play one or another of the, as I recall, three pinball machines. He was an excellent player. He knew just how much he could "jar" the machine during transit of a ball without getting a "tilt" from the machine. His ability with the flippers was extraordinary, repeatedly sending a given ball at the right speed in the optimal direction. He knew when a ball was headed for the exit place, and he often was able to jar the machine sufficiently to be able to save the ball and put it back in play. Most people there were amazed at his abilities.

An excellent example of Bob's sense of fair play and equality for all people occurred one night when Bob was at this local bar. The San Diego police came into the bar and arrested a person for some deed, an arrest which, to Bob, was totally unjustified. Members (cops) of police departments throughout America, particularly in large cities, are often "over zealous" in execution of their duties and the killing of people, particularly blacks and Hispanics, by police is well-known in recent times. Bob was horrified and very upset by the events of this arrest, both because there was no justification for the arrest and because of the tough manner in which the person was arrested and taken away. He talked about this extensively and repeatedly in the lab, and remained upset for weeks. He followed the case closely in the San Diego court system, and was active in pushing for a trial for the person rather than "settling out of court". When the trial came up, Bob testified for the defense. Bob's testimony was important, perhaps even crucial, to the subsequent acquittal of the person, with dropping of all charges. In so doing, Bob showed extraordinary courage and commitment, to testify in the court system against the local police, all in a foreign country far from home. Once again, Bob's open-mindedness and sense of equal treatment and rights for all came to the fore.

Being able to know and interact with Bob Pritchard was one of the real joys of my scientific career. He was bright, critical, alert, highly creative, and very knowledgeable. So many people with such qualities, in academic science as well as other professions, are also self-serving, egotistical, willing to do almost anything to get ahead and put others down. Bob was none of these things. He was truly interested only in finding the truth of what was going on, in getting to foolproof answers. His criticisms were directed only at alternative explanations to known results, to eliminating all ambiguities. In our laboratory, as the results began to delineate the conserved regions in the *E.coli* origin such as the DnaA binding sites and GATC sites and their methylation, discussions focused on the meaning of these sites for initiation of rounds of DNA replication, and such discussions were greatly enhanced by Bob's participation. Political discussions were also very rewarding and interesting. Bob would usually present a point of view more liberal and socialistic than most others, and it was always difficult to refute his argument. He was a delightful and truly imaginative person. He is sorely missed.

20 Bob Pritchard: Mentor, Teacher and Friend

Alfonso Jiménez-Sánchez*

I was introduced to the world of scientific research in Seville in 1970 by Enrique Cerdá-Olmedo with whom I obtained a PhD in molecular genetics. Afterwards, I discovered and enjoyed a new world of research at Stanford University in Phil Hanawalt's lab for two years. Both were certainly excellent mentors. They were the best until I went to Leicester University and met Bob Pritchard.

Throughout the seventies, my research was on mutagenesis at the replication fork in *E. coli* and on one of the ways that might open the tight coffer of the problem of thymineless death. My work was always related to chromosome replication but I never worked on replication itself, nevertheless I was very interested in papers related to this topic in the few journals received at the department in Seville and in the many in Stanford.

During that decade, a number of papers captured my imagination; surprisingly, many of them were written in the Department of Genetics at Leicester University and, among them, I would like to highlight those signed by Zaritsky and Pritchard (1970, 71, 73), and the most outstanding by Bob alone (1966, 68, 74). These papers made me want to go to Leicester and work with Bob.

---

*Universidad de Extremadura, Badajoz, Extremadura, Spain (a.jimenezsanchez@gmail.com).

Elena Guzmán and Bob Pritchard, **Segovia, 1987**.

In February 1981 I wrote a letter to Bob to show him how much I desired to go to his lab. It was the first of a number of letters written to prepare my visit. Yes, we wrote letters on paper, it was at the old pre-Internet age. That age had its own problems and advantages: it required a long wait for the answer to a letter but it was easier to keep letters. I have kept thirty letters from Bob, some of them handwritten.

I asked him for a proposal to apply for a fellowship. He had a very clear idea of what he wanted to do: "look for thermosensitive over-initiators and clone the replication inhibitor gene" (letter 27-5-1982). With this idea, I wrote the proposal and obtained a fellowship from the March Foundation for one year. As I had a lot of teaching duties at my University in Spain, I obtained the approval to split that period into two stays of six months. At that time I had a girlfriend, Elena Guzmán (now my wife and mother of two beautiful girls) who had recently finished her studies in Biology and was very interested in joining the project. On July 7th, 1982, after receiving Bob's agreement to bring her on board, we set out from Córdoba by car to arrive in Leicester three days later.

The first day in Leicester we went to the Department of Genetics. On entering, we saw a picture on the notice board with all the people working in the department. We got our first look of Bob. In our correspondence, Bob had told me that we were to work with Elisha Orr. As Elisa is a common name for girls in Spain, we thought Eli was a woman. The picture over the name of Eli Orr was a shock for us, everybody who knew Eli will know why (the same day I began writing this manuscript I was informed that he had recently passed away. He was messy but a hard worker, a very helpful colleague and a great friend). The first person we met was Margaret, the secretary. That was the second shock. I scarcely understood what she was trying to tell us (after one year of hearing her local accent I was still hardly able to make out what she said).

After interpreting Margaret's instructions, we found Eli, who came with us to look for Bob. Our first impression of Bob was of a gentleman. He was the head of a big department but he himself took care of helping us to find a place to live. Only after he was certain that we were accommodated did he decide it was time to talk about work. This was something Elena and I appreciated very much. At the beginning, we found in Bob a great mentor, soon he became our best teacher, later on we found a great friend.

During the first six-month period, we focused on obtaining *E. coli* mutants by EMS mutagenesis. They were selected by their thermosensitive colony-forming capacity and by quantitative colony hybridization using *oriC* sequence as a probe. The mutated genes of those with the best phenotypes were then localized on the *E. coli* map.

We worked very hard. The proof of this is that we tested 150,000 bacterial colonies, which let us isolate 16 mutants. We named them JG41 to JG56 and grouped them into two classes according to their phenotypes. Almost every day Bob appeared in the lab when he arrived in the morning and when he left in the late afternoon to ask the same question: "what's new?" as the preamble to talk about anything.

From our conversations with Bob, both Elena and I remember very clearly a day when we showed him a graph of the incorporation of radioactive thymidine by bacteria (as acid insoluble material) treated in two different ways. In this graph, both cultures showed exponential accumulation as two straight lines with different slopes. We interpreted the different slopes as corresponding to two different rates of DNA synthesis but Bob said "no my friends, they have a different explanation". He then

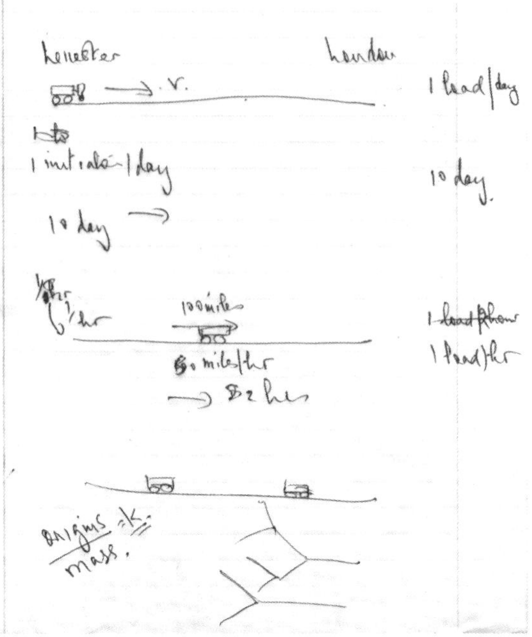

Fig. 1. Bob Pritchard's drawings, made in 1982, to illustrate his explanations to Elena and Alfonso about the comparison of a delivery truck with the replication process.

began to give us the best lesson we have ever received. I later realised that we were not the first to have been given that lesson but, most likely, we were the last.

Bob explained that if a company based in Leicester has to deliver one truck of their products in London every day, they have to arrange that one truck must leave from Leicester every day. One truck would then arrive in London every day irrespective of how long the trucks took to get from Leicester to London. This example made it perfectly clear that the rate of replication — like the speed of the trucks — was not what we were measuring but rather the frequency of initiation (corresponding to the frequency of trucks leaving Leicester). Since then we have repeated this simile every year in our teaching and, thanks to Elena's habit of keeping the drawings made by someone to illustrate some explanation, here I can show Bob's drawings when comparing deliveries with replication (Fig. 1).

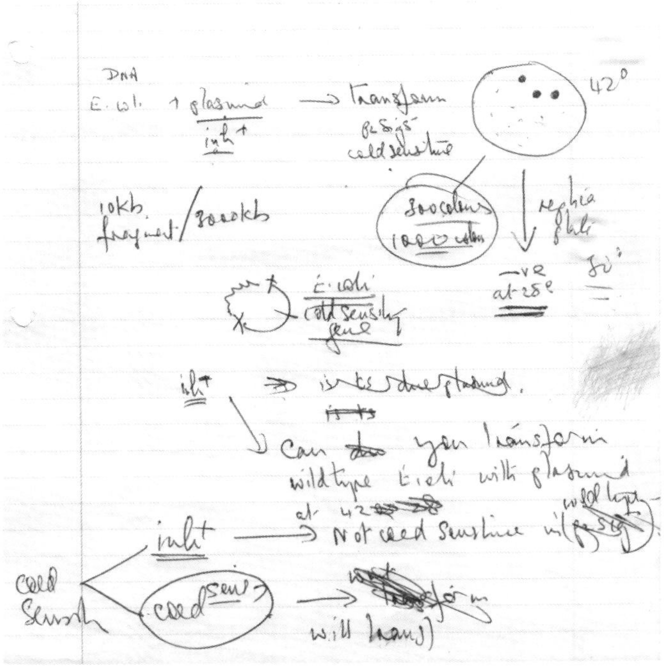

Fig. 2. Bob Pritchard's drawings, made in 1982, to illustrate his explanations to Elena and Alfonso about the plan for cloning the replication inhibitor gene.

One day, he may have felt that we regretted spending most of our time working in the lab so, to cheer us up, he said "Leicester is the best place to work, there is nothing to do or to see here".

By the end of our time in Leicester we had our work ready for publication. We showed that none of our 16 mutants had a mutation in a replication control gene and that what we had obtained were two kinds of mutants with mutations in either the *rpoB* or the *rpoC* gene, which code for the *b* and *b′* subunits of the RNA polymerase, respectively. We also showed that any wild type strain grown with a low concentration of rifampicin accurately mimics the phenotype of our mutants. Since a general, weak inhibition of the RNA polymerase generated the same phenotype as our mutants, this meant that their phenotypes were not the result of an alteration to some special activity of the RNA polymerase related to the control of the initiation of replication, but the result of a more general consequence of limiting RNA polymerase activity.[1]

We were not lucky with the mutation approach as we didn't find a control gene. The same happened with the cloning approach. The aim was to clone a chromosome fragment from a gene bank into an *E. coli* wild type strain and into the same strain lysogenic for phage P2 sig5, known to suppress inactive *oriC* only at high temperature. The cloning of the replication inhibitor gene will prevent any growth of the wild type strain, but will render the P2 lysogenic strain cold-sensitive. Again, Elena's habit lets me share with readers Bob's diagram of the cloning plan (Fig. 2).

After preparing two gene banks, we couldn't find any randomly cloned, chromosomal fragment that, when used to transform the strain with the integrated col-sensitive phage, would prevent the growth of the bacteria at low temperature. We know that Bob was very hopeful this approach would succeed; perhaps he was even aware that we were his last opportunity to get some evidence for the existence of the inhibitor gene. We couldn't. Could it be that there is not such a gene?

We finished our lab work in Leicester at the end of December 1983. By then, Bob had resigned from the headship of the Department of Genetics in order to spend more time on his other long-lasting passion, politics. We came back to Leicester in Easter of 1984 to finish the writing of our first paper together. When leaving Leicester we never said goodbye to Bob but "we'll see you soon".

We saw Bob again in the EMBO meetings held in Segovia (Spain, 1987), Collonges-la Rouge (France, 1990), and Sandhamn (Sweden, 1994). The French workshop was preparing the publication of a number of papers from the meeting and offered us to publish the work. I had presented at the end of the meeting. We had found that in the *E. coli* strain LE234, isoleucine starvation as opposed to starvation for the other amino acids, didn't inhibit initiation of replication. We know that the initiation of replication requires protein synthesis so this result points to the presence of a single protein devoid of isoleucine, the synthesis of which is the only protein synthesis required for initiation. We tentatively named this protein PinO.[2] Bob was very excited about these results and contributed with the analysis of the *E. coli* chromosome sequence (at this time, incomplete) to find the putative open reading frame. That was the last time we saw him.

Collonges-la Rouge, 1990.
1st row: Rolf Bernander, Lucien Caro.
2nd row: Elio Schaechter, Bob Pritchard, Kurt Nordström.
3rd row: X, Elena Guzmán, Alfonso Jiménez-Sánchez.
4th row: X, Eric Boye, X.

The achievements of anyone continue after life in his or her publications. Bob Pritchard left us a number of great publications. He will live among us for many years as long as someone reads them, talks about his teaching, or reads or writes about him. At this very moment he is living in our minds.

Acknowledgments

I thank Elena Guzmán for keeping and sharing with us the diagrams Bob used to explain to us his ideas. I also thank Vic Norris for his valuable corrections.

References

1. Guzmán EC, Jiménez-Sánchez A, Orr E, Pritchard RH. (1988) Heat stress in the presence of low RNA polymerase activity increases chromosome copy number of *Escherichia coli*. *Mol Gen Genet* **212**: 203–206.
2. Guzmán EC, Pritchard RH, Jiménez-Sánchez A. (1991) A calcium-binding protein that may be required for the initiation of chromosome replication in *Escherichia coli*. *Res Microbiol* **142**: 137–140.

21 Bob's Open House

Michael J. Pocklington*

I came to Leicester in the summer of 1982 for a job interview with Eli Orr in the Department of Genetics. I was pleased that the interview was going well, but even more pleased when someone asked me if I happened to have my swimming trunks. "We are off to the pool!" (Later I would discover that swimming trunks were not always obligatory). The interview was adjourned, and we squeezed into the departmental Land Rover, stopping on the way to pick up some beer. On the patio around the pool, the interview resumed, and Eli explained that on hot days in the summer, Bob Pritchard held an open house. Departmental members, and miscellaneous hangers-on, spent their lunch hour around the pool. Eli said that alongside him, under Bob, I would have total freedom to do whatever I wanted, provided I could defend my ideas with sound arguments. Grant money was pooled and everyone was allowed freedom to experiment, providing that the rationale could be convincingly made. A few beers later and my future in Leicester was settled.

Like Bob, Eli was a free spirit. Love of demonstrable scientific argument mattered above everything else. Any tradition or "received" wisdom was nonsense; it had no material (i.e. experimental) grounding; it had no authority. I do not know the extent to which Eli had got this from Bob, or

*6, Blankley Drive, LE2 2DE, UK (m.pock@me.com).

whether Bob had simply drawn Eli into his orbit because he recognised in him a fellow traveller; either way, Bob was the nucleus for a community of free-thinking people that included Eli. But this was no bunch of slackers: soon I would be participating in animated discussions that could go on into the early hours. An infectious sense of fun, freedom and irreverence prevailed, and it was probably responsible for the acknowledged success of the department.

Clearly a liberal in the true sense of the word, no subject matter was considered taboo by Bob. I recall having discussions with Bob where nothing was off-limits, even recreational drugs and bizarre sexual practices (other people's!). There was no such thing as a silly question and nothing to hold us back from discovering answers. You could apply practical scientific reasoning to anything, and back it up by experimental verification. This liberal ambiance was not an excuse for bad behaviour or sloppy thinking. As much as Bob loved to ridicule conservative nonsense, he also loved to ridicule fashionable nonsense. He could easily switch "sides" in an argument. You had to win him round using reason, or follow him to a novel conclusion. Of course, the whole idea of "sides" was absurd in the free and open marketplace of ideas that Bob espoused. Bob loved to be proven "wrong". He lived for the "Aha!" moment. We all did.

Bob's liberal temperament meant freedom for junior colleagues like me. What you did was your business, and also your responsibility. Although never judgmental or authoritarian, he could be mildly sarcastic. If you presented him with an idea that was getting too far ahead of itself, he might have said, with a wry smile, "good luck with that". He loved to puncture overblown rhetoric with a short no-nonsense question. Once, at the end of a seminar, he asked the student presenter a basic question. The student ended her discursive answer with the Australian questioning intonation, which was once fashionable, and is now ubiquitous. "Are you asking me or telling me?", Bob flatly inquired of the hapless student.

There was a time when I saw a lot of Bob, perhaps because my work was at a crossroads and I had time to spare. I used to enjoy mowing his lawn, and I helped him encourage wildlife. I used to clean his pool and rescue the frogs that had fallen in. The lunchtimes stretched into afternoons, and then evenings into nights. There were a lot of "Aha!" moments and we laughed. Bob identified the stars and we waited for meteors.

I remember one evening, relaxing on the patio, I was discussing with Bob my enthusiasm for obtaining mutants by selecting for compensatory mutations. When we force a genetic change, there is invariably a phenotypic effect selecting for a compensatory change somewhere else. It was like mutants resistant to antibiotics. The phenotype of a resistant mutant would show a growth disadvantage when the antibiotic was withdrawn, which Bob thought should be central to drug management in hospitals. Similarly, Bob thought, for this reason, that we need not worry too much about genetically manipulated genes: without artificial maintenance, they cannot compete against natural genes. It is like an abandoned flower garden which reverts to a refuge for wildlife. Bob said, "The weeds take over".

Looking for compensatory mutants was a good way to find new genes, such as those involved in the control of replication, cell division, and morphology. I liked the idea that each mutant could be purified and used as a basis for a new round of selection. In principle, we could "walk" along a genetic pathway, using repeated rounds of compensatory mutant selection and purification. We could imagine building a map of the genetic compensation network extending throughout the whole organism. I told Bob this was how evolution itself should be thought of. An organism was a naturally selected focus of interacting functions, like a neural network occupying an ecological niche.

Bob loved this vision because it was entirely "grounded" by genes and environmental factors. It typified the "genetics of function" approach, which he liked, as opposed to the "genetics of transmission", which exasperated him. It resonated with his vision of pragmatic liberalism. The compensation network was a metaphor of the real structure underlying society, constructed by the "invisible hand of the market"; and perhaps also of the mind, constructed by reward and punishment. Attempts at control inevitably lead to unintended compensatory consequences. Visualise this real structure (not an abstract formalism of it), and perhaps you could "nudge" it in the desired direction. I said I thought the compensation network might be more than a metaphor; it might be the structure underlying each and every evolutionary entity; a causal plexus within the wider causal nexus. It could form the basis for the experimental computer instantiation of consciousness. Bob wished me luck with that. But he was keen on the possibility of the worldwide monitoring of such things as antibiotic-resistant bacteria, and of technological, cultural, and economic trends. They would be like evolutionary weather reports. He acknowledged that the underlying structure would be something like the genetic compensation network. He would have been pleased, but perhaps impatient, with the now flourishing subject areas of antibiotic stewardship, systems biology, and network science.

In the lab we played with a lot of different ideas. Often we did not publish because it was more exciting to play. Under Barry Holland, who was another of Bob's innovative core, Eli established a large, diverse and successful yeast group. Later, a less progressive, medically-oriented regime assumed control. Eli clashed with the university authorities and moved his interests abroad. He continued to find scientific success, notably with monoclonal antibodies as therapeutic agents. Eli died in December 2015, the same year as Bob.

During the time I knew Bob, his interest in politics was growing. He seemed to enjoy local popularity. Shortly before the illness that incapacitated him, he came to my front door while canvassing my area for the local elections. I tried to persuade him that science was much more important than politics; that humanity could not save us, but science and technology would. He smiled and wished me good luck in my conviction. We each felt we had a duty to perform, but we were now on different trajectories. I suspect he knew he was entering the twilight, but he could not have known exactly what was in store.

My regular dog-walking route takes me past the vacant, overgrown plot where our pool parties once were. The pool I once cleaned has been colonised by algae and frogs; the lawn I once mowed, a temporary refuge for wildlife; on the patio where we once laughed and got intoxicated under the stars, a heap of rubble that was once Bob's house.

22 My Recollections of Bob Pritchard 1986–96

Vic Norris*

It was during my PhD in Dick D'Ari's lab in Paris, working on the coupling between DNA replication and cell division in *Escherichia coli*, that I began to appreciate the pioneering work of Bob Pritchard and his collaborators. Thanks to Dick, I learnt that the nature of the mechanism that regulates initiation of chromosome replication was one of several fundamental problems in the bacterial cell cycle. A friend in a nearby lab, Francoise Gosti, introduced me to calcium as a major player in the cell cycle of eukaryotes and this made me wonder whether calcium might play a similar role in prokaryotes. Calcium was known as a regulator of myosin, so when Barry Holland on a visit to Dick's lab told me that Eli Orr (who was also in the Genetics Department) had found a myosin in yeast, I saw that Leicester was the place to be. I applied for an EMBO award to join Barry and was interviewed in Uppsala by Kurt Nordstrom. Kurt, who knew the Genetics Department well, said that it was probably the best in the world. After the interview, Kurt told me about his own work with *int* strains in which chromosome replication depends on the integration of a plasmid origin of replication into the chromosome. My failure to see why this might be interesting was tactlessly all too evident. Nevertheless, Kurt supported my application and I went to Leicester, where I met Bob in the flesh.

*University of Rouen, 76821 Mont Saint Aignan, France (victor.norris@univ-rouen.fr).

Bob Pritchard, talking at his retirement party, 1989, Department's 25th Jubilee 'Feast'.

Bob immediately impressed me. Many of the results from his group, ranging from thymine limitation[1] to gyrase[2] to minichromosomes[2] were clearly relevant to my own interests and, in particular, a clever experiment that had led to the discovery of PinO, a putative calcium-binding protein that was a candidate for being involved in regulating initiation.[3,4] I was struck by Bob's intelligence every time we had a discussion, whether scientific or political. His analyses were both lucid and subtle. I was also struck by his wide-ranging curiosity. For example, running home in the evenings, I sometimes stopped off at his house on Knighton Grange Road where Bob revealed his interest in botany in explaining, for example, how the plants in his garden dispersed their seeds.

Another of Bob's many outstanding qualities was his honesty. I remember asking for his opinion on a letter I intended to send to one of the funding bodies. The next day, standing at the bar in the Charles Wilson building, he told me my letter was rather plaintive. He was absolutely right and, by opening my eyes, had done me a big favour. He himself, however, had taken a risk by being honest. I was therefore doubly grateful. From then on, I have tried to follow his example when asked by friends for my opinion.

Since returning to France, I have continued to think about the initiation of replication. This has led me to revisit several of Bob's papers, the significance of which I had failed to grasp when I was in Leicester. These papers include one that reveals a relationship between initiation, RNA polymerase activity and growth temperature.[5] For the last decade or so, I have been giving a course in Scientific English, which is based on one of Kurt Nordstrom's *int* papers[6] — life is full of irony. The results of this paper can be interpreted to mean that neither accumulation of an activator nor dilution of an inhibitor with respect to the chromosome can constitute the mechanism that times initiation.[7,8] In this course, we discuss attempts to clone the origin of chromosomal replication (e.g. Ref. 2) and the implications of the success of these attempts for models of initiation control, in particular, the ability of *E. coli* to replicate its chromosome despite the presence of scores of minichromosomes replicating in synchrony with it.[9–11] I draw the students' attention to the significance of the fact that these results led to Bob abandoning this type of model: Science advances not only when we get it right but also when we acknowledge we get it wrong. Bob's success in doing both is a measure of his remarkable calibre.

References

1. Zaritsky A, Pritchard RH. (1973) Changes in cell size and shape associated with changes in the replication time of the chromosome of *Escherichia coli*. *J Bacteriol* **114**: 824–837.
2. Diaz R, Pritchard RH. (1978) Cloning of replication origins from the *E. coli* K12 chromosome. *Nature* **275**: 561–564.
3. Guzman EC, Pritchard RH, Jimenez-Sanchez A. (1991) A calcium-binding protein that may be required for the initiation of chromosome replication in *Escherichia coli*. *Res Microbiol* **142**: 137–140.
4. Guzman EC, Jimenez-Sanchez A. (1991) Location of *pinO*, a new gene located between *tufA* and *rpsJ*, on the physical map of the *Escherichia coli* chromosome. *J Bacteriol* **173**: 7409.

5. Guzman EC, Jimenez-Sanchez A, Orr E, Pritchard RH. (1988) Heat stress in the presence of low RNA polymerase activity increases chromosome copy number of *Escherichia coli*. *Molec Gen Genet* **212**: 203–206.
6. Eliasson A, Nordstrom K. (1997) Replication of minichromosomes in a host in which chromosome replication is random. *Molec Microbiol* **23**: 1215–1220.
7. Donachie WD. (1968) Relationship between cell size and time of initiation of DNA replication. *Nature* **219**: 1077–1079.
8. Pritchard RH, Barth PT, Collins J. (1969) Control of DNA synthesis in bacteria. *Symp Soc Gen Microbiol* **19**: 263–297.
9. Yasuda S, Hirota Y. (1977) Cloning and mapping of the replication origin of *Escherichia coli*. *Proc Natl Acad Sci USA* **74**: 5458–5462.
10. Messer W, Bergmans HE, Meijer M, *et al.* (1978) Mini-chromosomes: plasmids which carry the *E. coli* replication origin. *Molec Gen Genet* **162**: 269–275.
11. Leonard AC, Helmstetter CE. (1986) Cell cycle-specific replication of *Escherichia coli* minichromosomes. *Proc Natl Acad Sci USA* **83**: 5101–5105.

23 Bob Pritchard and the Rookie Geneticist

Sir Alec J. Jeffreys FRS*

I count myself fortunate indeed to have known Bob — a person with one of the sharpest minds I have ever known, a brilliant scientist and above all a good and loyal personal friend. I first heard of Bob in the late 1960s as an undergraduate at the University of Oxford where his seminal work on the control of bacterial DNA replication was covered in one of the courses, work incidentally that paved the way to understanding cell cycle control resulting in a Nobel Prize for Harwell, Hunt and Nurse in 2001. One sad fact about Bob is that his work never received the recognition it was due, neither by the UK science establishment nor by the international community.

My first direct contact with Bob was in 1977 when I was a postdoc at the University of Amsterdam. I was keen to set up my own laboratory and was applying to UK universities for a lectureship. One application was to Leicester, and Bob phoned to say that they were interested and asked me to attend for interview. My first problem was locating Leicester — all I knew was that it was somewhere north of Bedford and south of the Scottish border, but fortunately one of my Dutch colleagues had a small Michelin map of Europe to hand which by good chance showed Leicester at the heart of England, which seemed as good a location as anywhere for my future.

*Department of Genetics, University of Leicester, Leicester, LE1 7RH, UK (ajj@leicester.ac.uk).

I was immediately impressed with the Department of Genetics at Leicester, and especially with the warm welcome from Bob, Barry Holland, Brian Wilkins, Graham Bulfield and colleagues. I immediately sensed that this was a place mercifully free of prima donnas where research could be fun. The position on offer was a temporary lectureship funded unusually not by the School of Biological Sciences, of which Genetics was a member, but instead by the relatively new School of Medicine. Back in those days, genetics was very much seen as an obscure fringe component of medicine, and it is a testament to the powers of persuasion of Bob and his colleagues that the Medical School was prepared to take genetics seriously, with a substantial component of genetics lectures, tutorials and practicals provided to undergraduate medics. Bob also saw that molecular genetics, just beginning to emerge, was going to be a transformative technology and that the Department needed to move into this field, hence their interest in me. At the time, the Department's focus was primarily on bacterial and fungal genetics, though population genetics and mouse genetics were already represented by Robert Semeonoff and Graham Bulfield respectively.

I arrived in the Department in September 1977 at the tender age of 27 and was presented with my own lab, a moment of joy that can hardly be described. OK, it was dull, dingy and nearly empty, but that didn't matter — this was my new kingdom. One of the first events was a meeting with Bob, who welcomed me to the Department, outlined my duties, then finished by saying "If you don't give me any trouble, I won't give you any". This was coded language for "if you get on with your research and have fun, I won't interfere". This was music to my ears, and Bob was as good as his word — he always encouraged and took a real interest in the work of his colleagues, but never ever interfered or tried to direct. Instead, he provided a place of total academic freedom that proved so important to my subsequent work that led to DNA fingerprinting. Indeed, the history of forensic DNA and my place in it might have been radically different were I not embedded in such a laissez-faire research environment.

Bob was keen to see the new technology of molecular genetics and genomics introduced into the Department, and awarded me with Jim Mackley as a half-time technician. My initial research goals, helped greatly by Bob's implementation of pooled funding that gave me access to money even without a grant, were to explore human gene family organization and evolution, and also to see whether we could detect sequence variation directly in human DNA. We soon had the research tools of the time (DNA extraction, restriction enzymes, probe preparation, Southern blotting etc.) up and running, and I was delighted to see how quickly this spread through the Department. We published our first description of human RFLPs in 1979 and then went on to identify hypervariable DNA sequences in the human genome, work that led eventually (and accidentally) to the first DNA fingerprints in 1984. Bob was intrigued by the practical applications of our work, I suspect primarily through his interest in politics — he could see as well as I the potential impact on police practice, immigration control, family law, the courts and the like.

I also threw myself into the world of lecturing, with my first lecture given to first year medics, a true baptism of fire. Again, Bob never interfered but just kept a gentle eye on me through his colleagues to make sure that I was not making a mess of things. On one occasion, Bob asked whether I could take on two of his lectures as he was otherwise committed. I was very honoured and took on the task despite knowing very little about the subject matter (it included tRNAs as I recall). Bob thanked me profusely afterwards and rather sheepishly offered me an apple from his garden as a token of thanks — a simple but very moving gesture from a kind and thoughtful man. Bob's garden became very special to many of us; Bob was the proud owner of a huge house complete with swimming pool in the back garden (almost unheard of in Leicester), and he gave us free rein to use the pool whenever we wanted. It was wonderful to relax with colleagues by the pool on a hot summer's day and to solve all the problems of the world and especially to discuss where our science was going.

Bob's retirement from the University in 1983 to allow him to enter politics full-time was a great shock to us all. Bob treated us all as one large happy family, and we rightly saw his departure as akin to being orphaned. However, Barry Holland stepped into the breech as the new Head of Department, and pressed on with Bob's vision of a Department spanning the breadth of genetics from bacterial to human. The Department has gone from strength to strength, expanding greatly and developing an enviable reputation in bacterial genetics, human genetics and genomics, medical genetics, behavioural genetics, evolutionary and population genetics, and most recently in neurogenetics. It also has a very active outreach programme aimed at schools, interest groups and the lay public. It has been recognised by Queen's Awards in 2002 and 2013, and achieved international recognition for DNA fingerprinting and for the recent discovery of the mortal remains of King Richard III, the last of the Plantagenets, under a car park in Leicester. Bob would have been proud of this progress, all of which was rooted in his vision of molecular genetics back in the 1970s. His spirit still hovers over the Department, with its friendliness, collegiality and his communal budget system that survives to this day.

Bob's move into politics as a Liberal councillor marked another major chapter in his life. My own somewhat cynical view was that he was far too bright for politics, and that he probably was not really cut out for pressing the flesh and baby kissing. In any event, his attempt to win a parliamentary seat failed and he resigned himself to local politics. Our paths still crossed, most notably over plans to build a ring road past our house. The Council was in favour of a very large dual carriageway, while Bob pressed for dropping this into a cutting to reduce noise levels though this would have cut our community in two. In collaboration with a local residents' committee, I took on the task of drawing up an alternative plan that would scale down the road and keep the community intact. The Council accepted this and the final road plan was very much as I envisaged. This was my first and only foray into road planning, and the only time I got the better of Bob. As ever, he accepted defeat with a wry smile and very good grace.

It has been a privilege to work for very nearly the whole of my career in the Department that Bob founded, and my move to Leicester was the best career decision of my life. But maybe it was predestined. Some years ago, I appeared on Desert Island Discs, a BBC radio programme where people talk about their lives against a background of eight pieces of music that they would like to take if stranded on a desert island (my choices were aimed at annoying the listeners, apparently successfully too). During the chat, I recounted a story about my paternal grandfather and was intrigued to hear from someone after the programme who told me that my description of my grandfather matched that of one of their uncles. To cut a long story short, it turned out that this person was a first cousin of my father from a branch of the family that had split off and been forgotten by us many decades ago. Two further unknown first cousins of my father subsequently emerged, all from this lost side of the family. My father was a keen watercolour painter

Complete reunion of the participants of the 50th Anniversary of the department, in front of the Adrian Biology Building, October 11th, 2014. (*Too numerous to be enlisted by name.*)

(Photograph courtesy of Department of Genetics, University of Leicester.)

and dinghy sailor, so it was most intriguing to learn that one of the lost cousins was also a highly accomplished watercolour artist and that another was also a dinghy sailor who had even competed against my father without knowing that they were related. The third unknown cousin was the most intriguing — she was a retired microbiologist strongly associated with the Lister Institute of Preventive Medicine, of which I was once a research fellow and am now a Governor. And with whom did she collaborate closely back in her Lister days? None other than Bob, the very man who appointed me to Leicester. It seems as if the Jeffreys' destiny is written in our genes, with one gene ensuring collaboration with Bob. All very spooky!

24 The Last Thirty Years: Bob Pritchard's Legacy to Genetics and Society

Peter A. Meacock* and Annette Cashmore†

The Department of Genetics that Bob Pritchard founded continues to be one of the most dynamic and outward-looking departments at the University of Leicester, yet it still retains the philosophy of collegiality, collaboration and mutual support that he instilled from its foundation in 1964. Here we have attempted to record some of the major achievements and prime movers of the Department's development since Bob Pritchard relinquished the Headship.

In the 1970s, astute judgement by Pritchard and the Heads of the other biological sciences departments about the directions of future biological research led to recruitment of new staff practised in modern molecular technologies. As a result Leicester became one of the earliest of UK Universities to have a strong interactive pool of staff with expertise in the then new Recombinant DNA technologies for gene cloning and analysis. As well as in Genetics (Alec Jeffreys, Barry Holland, Brian Wilkins, Peter Williams, Bob Pritchard, Eli Orr), other strong groupings existed in Biochemistry (Bill Brammar, Bill Shaw, Eric Blair), the ICI-University Joint Laboratory directed by Bill Brammar (Peter Meacock, John Windass) and Microbiology (Jeff Almond).

*Department of Genetics, University of Leicester, LE1 7RH, UK (mea@le.ac.uk).
†Department of Genetics, University of Leicester, LE1 7RH, UK (deceased, 28-05-2016).

Barry Holland, who eventually succeeded as Head of Department after Clive Roberts, recognised Leicester's unique situation and was the driving force behind two significant developments which had major impact. One, from 1981, was the organisation and delivery of highly intensive laboratory-based courses in "Gene Cloning and Analysis" for scientists from academia and industry, with financial support for UK academic researchers from the Biotechnology Directorate of the UK Science and Engineering Research Council. Over two intensive weeks, participants had hands-on training in "state-of-the-art" techniques of cDNA cloning in plasmid vectors, mammalian genomic DNA library construction using bacteriophage *Lambda* vectors, and both *in vitro* and *in vivo* analysis of gene structure and expression. The courses, which took participants from Universities and companies in the UK, Europe and further afield, established Leicester as a leading centre of research and teaching based on these advanced methodologies.

The success of the Leicester cloning courses led to the department offering a one year research-orientated MSc Programme in Molecular Genetics, which was first run in 1987 with Clive Roberts as Convenor, and subsequently for several years by Eli Orr, Julian Ketley and Mike Hennessey before the present incumbent Fred Tata. That course has now been running for almost 30 years, and still continues to recruit well, especially from the international student sector.

The second achievement of Barry Holland was the establishment of the Leicester Biocentre in 1982. Realising the keen interest from companies in the biotechnological potential of molecular genetics, and the successful one-on-one model of the ICI-University Joint Research Laboratory founded by Prof. Bill Brammar of Biochemistry, Barry had the vision to establish a new type of research centre at Leicester — an academically-orientated centre carrying out fundamental core molecular genetics research in areas of interest to industry, and offering companies the chance to place their own scientists alongside the academics to work on projects of the company's choice. Thus, companies could carry out their own research in a supportive academic environment drawing upon the expertise and facilities available there. So the Leicester Biocentre was born linked to but separate from, the Department of Genetics, with sponsoring companies Whitbread, Dalgety, Distillers Yeast Co, John Brown Engineers and Gallaher Tobacco. Because the venture was building links between UK companies and academia, the Biotechnology Directorate of the UK Science and Engineering Research Council awarded a grant for capital equipment support, and the Wolfson Foundation provided funding for a new purpose-designed building to house this exciting new venture, one of the first university-based biotechnology centres in the UK.

The research fields, selected by Barry Holland and the sponsoring companies, focussed on yeast and plants with the initial Research Group Leaders as Alan Boyd and Peter Meacock, both Leicester Genetics PhD graduates, and Rick Walden. Further research income was won through grants from UK Research Councils, charities and government agencies, and through internationally funded contract research. The Leicester Biocentre functioned successfully as an independent self-supporting research centre providing service to its sponsoring companies for 10 years until 1992; by then the sponsors had decided either to establish molecular biology as part of the company's in-house research programme or to diversify into other commercial directions not needing genetics research. The remaining yeast and plant core research was moved into the Departments of Genetics and Biology, and Genetics took over usage of the Biocentre building to supplement its fully occupied accommodation in the Adrian building.

The opening of the Medical School in 1975, and further expansion in the 1980s, led to recruitment of new academic staff in order to carry the extra teaching load; they included Raymond Dalgleish with expertise in human molecular genetics relating to connective tissue disorders, and Charalambos (Bambos) Kyriacou, who brought new expertise in behavioural genetics, particularly on biological clocks and biorhythms of the fruit fly *Drosophila melanogaster*.

1984 was a momentous year for the Department. From fundamental research on mammalian gene structure and genome variation caused by mini-satellite DNA sequences, Alec Jeffreys published his paper in the periodical *Nature* that announced to the world the development of "DNA-fingerprinting" and its use in determining inheritance relationships. The implications of being able to identify human individuals from molecular analysis of their DNA were recognised as ground-breaking and quickly applied to a range of practical uses including forensic investigations (the Enderby murders), immigration paternity disputes and the identification of mortal remains of important historic individuals, notably at that time the Russian Royal Family and the Nazi mass exterminator Josef Mengele. Later, in 1996, using Alec Jeffreys' methodology to examine the DNA of survivors from the Chernobyl nuclear disaster, Prof. Yuri Dubrova proved unambiguously that nuclear radiation causes germ-line mutations in humans. Moreover the technology could be applied to other animals, and even plants, to aid in breeding and conservation programmes, and DNA-fingerprinting was used to prove unambiguously the parentage of the first cloned animal, Dolly the sheep. In recognition of the importance of his scientific contributions Alec received numerous awards including a Knighthood in 1994, Freedom of the City of Leicester in 1992 and many prestigious scientific awards from institutions worldwide, bringing the Leicester Department of Genetics to even greater prominence.

1989 saw the Department celebrating its 25th Anniversary with an international Symposium on "Genetics and Society" organised by Barry Holland and Bambos Kyriacou as the first large-scale venture into public engagement on genetic research. The event was open to the general public with Lord George Porter, the University Chancellor and President of the Royal Society, acting as Chairman to introduce a list of eminent expert speakers who included Baroness Mary Warnock (Philosopher, Ethicist and Chair of the House of Commons enquiry that led to the Human Fertilisation and Embryology Act of 1990), Dr. Jim Watson (Nobel Laureate for the discovery of the structure of DNA, Director of the Cold Spring Harbor Laboratory, New York, and Coordinator of the Human Genome Initiative), Prof. Tom Caskey (Advisor on the Human Genome Project and Director of the Institute for Molecular Genetics at the Baylor College of Medicine in Houston, Texas), Dr. Anne McLaren (Director of the MRC Mammalian Development Unit), Prof. Richard Flavell (Plant molecular geneticist and Director of the John Innes Institute at Norwich) and Leicester's own Bob Pritchard and Prof. Alec Jeffreys (developer of DNA-fingerprinting). The public responded with great enthusiasm at the chance to hear how modern genetic research was impacting on society, and almost 2000 people travelled from far and wide so that the Leicester De Montfort Hall was packed for the occasion. Eventually, a book (Genetics and Society; Addison-Wesley Publ. Co, 1993) was published of the proceedings.

After Bob Pritchard, Headship transferred successively to Clive Roberts, Barry Holland and then Peter Williams who each had their own style of running the department. In 1992, linked with the transfer of Peter Williams to lead the new Department of Microbiology, the University decided to recruit externally for a new Head of Genetics rather than promote from within its own ranks. The new appointee was Prof. Gabriel (Gabby) Dover from the Department of Genetics at Cambridge, with expertise and interests in the molecular processes and drivers of evolution. This dimension of Genetics had not previously been a major strength of the department but, in view of the advances going on in science at that time, was becoming central to the teaching and understanding of the discipline. Dover also brought alternative ways of looking at the department's teaching portfolio, and challenged the whole School of Biological Sciences to readdress its undergraduate curriculum, leading to the introduction of subject-named specialist degree streams; previously undergraduates had simply taken a BSc in "Biological Sciences".

Importantly for the department, the move of Dover to Leicester also attracted Mark Jobling from Oxford as a Welcome Trust Research Fellow; he studied variation and inheritance of the maleness-determining Y-chromosome in order to investigate human evolution. Jobling's research, particularly in association with his ex-PhD student, now Lecturer, Turi King, has become one of the most high-profile research areas and continues the media-interest in the department triggered by Alec Jeffreys' work. Research projects that have received acclaim in the scientific and national media have included:- analysis of the linked inheritance of surnames (family names) in western society to passage of the Y-chromosome through generations; use of Y-chromosome inheritance to study the spread of human populations, notably the Vikings in Britain, and demonstration of a predominantly Neolithic origin for European paternal ancestry; examination of the ancestry and descendants of Thomas Jefferson, principal author of the Declaration of American Independence and third President of the USA.

After Dover completed his tenure as Head in 1995/6, responsibility for leading the department was taken up by Brian Wilkins. There was a major refurbishment of the research laboratories so that the expanded department, which had previously had to make temporary use of the Biocentre Building, could all be housed together in the Adrian Building.

These moves consolidated the department so that it was better able to focus on its prime missions of teaching and research.

The 1990s saw the launch of a new set of undergraduate Biological BSc degrees which were designed to prepare students for careers in non-clinical medical sciences; so Genetics offered a Medical Genetics programme alongside its conventional Biological Sciences (Genetics) degree. Much of the teaching was delivered by new staff who held joint appointments between Genetics in the School of Biological Sciences and a department of the School of Medicine; in particular Prof. Richard Trembath (joint with Cardiovascular Sciences) and Prof. Paul Burton (joint with Genetic Epidemiology/Health Sciences). Delivery of the Medical Genetics degree also led to the formalisation of links with staff of the NHS Genetics Counselling Centre at the Leicester Royal Infirmary who also contributed to the teaching.

During this time, Peter Meacock rejoined the Department from the Biocentre, and with Eli Orr established the area of yeast genetics in the Department; on his advice and guidance, the Department recruited

The Adrian Building, home of the Department of Genetics since 1967. From initially being housed in less than one half of the First Floor, the department has expanded to now occupy the whole building. (Photograph courtesy of P.A. Meacock.)

the yeast geneticists Ed Louis and Rhona Borts from Oxford. Ed Louis was at the forefront of research into the molecular structure and processes at telomeres of eukaryotic chromosomes, complementing the work of Nicola Royle on human telomeres, and Rhona Borts was studying the role of DNA repair mechanisms in the process of meiosis, which brings about the production of gametes in eukaryotes.

Brian Wilkins was succeeded as Head for an interim 18 month period by Bambos Kyriacou, whose own research on biological clocks was by then receiving much international recognition, specifically to coordinate the departmental submission to the forthcoming national Research Assessment Exercise (RAE); that submission led by Kyriacou and based on Wilkins' consolidation of the department resulted in Genetics receiving the highest possible RAE score, making it the only 5* university Genetics department in the UK.

In 2000, Annette Cashmore took over as Head of Department. She had originally come to Leicester in 1983 as a postdoc researching the yeast *Saccharomyces cerevisiae* with Peter Meacock in the Leicester Biocentre and subsequently secured a Lectureship in Genetics, linked to Alec Jeffreys' Lister Fellowship, where she established her own research on the human fungal pathogen *Candida albicans* investigating the role of iron uptake and metabolism in pathogenesis. One of her first jobs as Head of Department was to receive the exciting news about the RAE 5* rating which led to a period when the University looked kindly on Genetics, so making the role of Head enjoyable! In the five years that Annette was Head, the Department grew from 14 to 28 academics, each with their own research group. Tony Brookes and Richard Badge brought a new dimension to the Department, focussing on bioinformatics and the developing new approaches to sequencing and analysing genomes. Flaviano Giorgini came with interests in human neurological conditions such as Huntington's and Alzheimer's diseases, and using yeast as a model system to identify the causative genes.

Celia May, Ed Hollox and Chris Talbot all joined the department at this time, increasing the size of the Human/Medical Genetics Group. Also Raffael Schaffrath, who had been a PhD student with Peter Meacock, returned to Leicester to add to the yeast genetics activity. Peter Williams came back to the Department bringing with him Colin Hewitt and Julie

Morrissey, and they were joined by Chris Bayliss from Nottingham. This influx greatly extended the Microbial Pathogenicity Group.

It was also during this time that Eran Tauber and Ezio Rosato were appointed as Lecturers, so joining their mentor Bambos Kyriacou to establish a large and dynamic "fly" group with research addressing biorhythms and behaviour. The research of this group has been a major contributing factor in the successful development of Neuroscience at Leicester. Today the three fly groups led by Kyriacou, Rosato and Tauber occupy the entire space that housed the whole Genetics department in 1971.

As well as having a comfortable relationship with the University following the high RAE rating, the Department itself was easy and delightful to lead because the mutually supportive environment meant that new things could be tried and people's enthusiasm relied on. This stemmed from the environment that Bob had established right from the beginning of the department. Bob's advice to all subsequent Heads was to always "make sure that everyone knows what is going on and has opportunity to voice their view about any decisions being made, but at the end of the day do what you (the Head) feels is best for the department"; this was the politician in him!

In 2002, the Leicester Institute of Genetics and Genomics was established. This was a "virtual Institute" as it did not have a building, but it brought together people from around the University who were using genetics as part of their research. The Institute was formally opened by Sir Paul Nurse, after a British Association for the Advancement of Science meeting at Leicester, where Sir Harry Kroto had presented Sir Alec Jeffreys with a Royal Society of Chemistry award recognising the Adrian Building as a "Chemical Landmark" site, for the discovery of DNA fingerprinting. That day Annette Cashmore was chairing events and shepherding three Knights of the Realm, none of whom wore formal ties, nor in one case even a collar on his shirt, and all were maverick in what they did and said — never sticking to any agenda! However, it was a day of great fun and enthusiasm, and Bob's influence was apparent in how collegiality, passion and a love of the subject, were the driver for getting the Department at the forefront of everything.

2002 also saw another exciting episode through the award of the Queen's Anniversary Prize. These prizes are awarded every two years to a few institutions for projects considered to have wide impact and reach.

Entrance Foyer to the Adrian Building. In 2002, the building was designated as a "Chemical Landmark" site by the Royal Society of Chemistry to mark Alec Jeffreys' discovery of DNA-fingerprinting. The DNA model comprises two repeats of the first "mini-satellite" sequence used for DNA-fingerprint analysis. The plaque itself is mounted on the outside of the building beside the main entrance doorway. (Photograph courtesy of P.A. Meacock.)

With "encouragement" of the then University Vice-Chancellor, Prof. Sir Bob Burgess, the whole Department was put forward under the theme "Genetics, Science and Society". The proposal demonstrated how the synergy between teaching and research in the Department had huge impact not only on the advancement of science, but also on increasing public understanding and engagement both locally and world-wide. Winning the prize meant that Annette Cashmore, as Head, went to Buckingham Palace with the VC to receive the plaque and certificate from HM the Queen and Prince Philip. Other members of the Department also went to either the celebration dinner at the London Guildhall or to Buckingham Palace to see the receiving of the prize. Several Genetics undergraduate students were also invited and everyone was to be introduced to HM the Queen by Sir Michael Atiyah, the University Chancellor. Before the event, the Chancellor had come along to the Department to meet the students and to make sure he had got everybody's name right. To Annette's horror on the day, he introduced her to the Queen and asked her to introduce everyone else. Unprepared, she did not know how to pronounce all of the student's names, but in true Genetics style they "went with it" and the Genetics delegation came out looking "together and professional"!

Another venture that the VC, Bob Burgess, "encouraged" the department to do, in order to show off its genetics excellence, was to apply for a Centre of Excellence in Teaching and Learning (CETL) award. The Labour Government had made available £315 million to establish 74 such centres throughout England. In 2005, the bid was successful and Leicester Genetics was awarded £5 million to establish a CETL called GENIE (Genetics Education Networking for Innovation and Excellence) led by Annette

Another (Double) Anniversary — Another Cake! From *left* to *right*: Annette Cashmore, Bambos Kyriacou (with son), Peter Williams, Sir Alec Jeffreys and Barry Holland in 2004, cutting the cake to celebrate the 40th Anniversary of the Department and the 20th Anniversary of DNA-Fingerprinting. (Photograph courtesy of Department of Genetics, University of Leicester.)

Cashmore. GENIE had four strands of work: firstly, capital funding for improvements to Genetics teaching facilities — laboratories, computer and small group teaching rooms, and other refurbishment of research space in the Adrian Building. These improvements, which included installation of PA and cctv to aid teaching in large laboratories and expansion of the department's space, were possible because of the Biochemistry Department's move to a new, purpose-built facility. Secondly, the improvement of both genetics-specific teaching and teaching approaches generally relevant to all aspects of higher education, and the development of projects bearing on higher education policy and strategy, for example on reward and recognition for teaching and learning. This led to the establishment in 2013 of a University-wide academic career stream from Junior Lecturer to Professor based on teaching-focused activities, wherein teaching was given parity with research. Thirdly, the establishment of a web-based Virtual Genetics Education Centre with support platforms for higher education, schools and colleges, health professionals and the general public, which now receives 4,000–5,000 hits per day from all around the world, i.e. "outreach" in its broadest sense! Fourthly, local outreach work to enhance public engagement in science, through hands-on "Dynamic DNA" days for school children and teachers, and presentations to local community groups. More than ten years later GENIE-CETL still exists and is now part-funded by the University. Bob Pritchard had always advocated that the general public should be educated about scientific developments so that they were better able to evaluate proposed policies of government and other decision-making bodies that would affect society, and that outreach was therefore an important function of a university department; no doubt he would have proud that his department was taking an active role in this type of venture.

The creation of GENIE brought three new academic staff to the department (Cas Kramer, Mark Goodwin, and Nicola Suter-Giorgini) who were initially appointed as Teaching Fellows, but later given full Lectureships under the new teaching-based career progression scheme. Other members of the department have become actively involved in GENIE's work including Chris Cane, Raymond Dalgleish, Eran Tauber and Julian Barwell (who joined the Department as a clinical geneticist succeeding Richard Trembath), and have won international awards for their innovative developments in teaching and its delivery. In addition, Annette Cashmore was awarded a Higher Education Academy (HEA) National Teaching Fellowship, and all members of GENIE have received HEA University Teaching Fellowships. Currently, 27 members of the department have now received HEA Teaching Fellowships which are now recognised as being higher education teaching qualifications, with help and mentoring from GENIE and support from Julian Ketley, the current Head of Department. The department's aim is to have all academic staff holding a HEA Fellowship, a clear indicator of the excellence of Genetics teaching at Leicester!

When Annette transferred formally to become Director of GENIE in 2005, Peter Williams took up his second term as Head of Department. He was followed in 2009 by Julian Ketley, a bacterial geneticist with specialist interests in *Campylobacter* pathogenesis, who is the current Head. Julian initially came to the department as a Royal Society University Research Fellow to work alongside Tim Hirst, a Wellcome Trust Fellow, Barry Holland and Peter Williams. He has since fulfilled many teaching roles while building a strong microbial pathogenesis research community within the department with close ties to the Department of Infection, Immunity & Inflammation in the Medical School. Since Julian has been Head, Ed Louis, a previous Professor in the department, has returned from a brief period at the University of Nottingham as Director of the new *Centre for Genetic Architecture of Complex Traits,* and Prof. Marco Oggioni has joined from the University of Siena to expand further the Microbial Pathogenicity group.

In 2014 Dr. Turi King, who had been a MSc student in the department and had completed her PhD with Mark Jobling, was appointed as Lecturer jointly with the School of Archaeology. As part of a University-wide multidisciplinary team Turi had been involved in the discovery of the mortal remains of the last English Plantagenet King, Richard III, from a grave under a Leicester car park and she had carried out the DNA analysis that proved his identity; a project that received world-wide media attention and much acclaim for both the University and the Department. She featured prominently in the media coverage when the king's remains were reinterred in Leicester Cathedral in March 2015. Other staff who arrived during the 2000 to 2014 period were Steven Foster and Sandra Beleza, both with interests in human genetics.

The department also featured on TV in 2015 when *Code of a Killer*, a television police drama covering the story of the use of Prof. Alec Jeffrey's DNA fingerprinting technology to catch the murderer of two Enderby schoolgirls, was shown on Channel 4. This had been preceded by visits of the camera crew to the department, and around the University campus, trying to recreate the 1980s when the DNA fingerprinting technology had been used both to establish the innocence of an original, self-confessed, suspect and to identify the real killer.

Throughout the department's history its staff have always made significant contributions to the wider academic community through their services on numerous University and Faculty, later College, committees. Bob Pritchard had been Chairman of the School of Biological Sciences, the body responsible for overseeing the design and teaching delivery of the undergraduate Biological Sciences degrees; other Heads followed in his footsteps — notably Peter Williams and Brian Wilkins. Annette has served on this committee for 20 years and currently there are four members of the department playing important leading roles in admissions and examinations. Similarly, several staff have served on the Academic Committee of the Medical School, and Annette Cashmore was Faculty Sub-Dean. Likewise, Peter Meacock was, for 10 years, Graduate Sub-Dean and Director of Postgraduate Research for the Faculty and College of Medicine, Biological Sciences & Psychology. Peter Meacock was a major driver for the formation of the University Graduate School, leading to change in the selection, admissions and training of postgraduate research students.

No great university department achieves its success purely from the efforts of its academic staff — an efficient, dedicated and well-managed team of support staff, who are regarded as equal contributors, has always been a strength of the Department of Genetics. Three figures who were recruited by Bob Pritchard in the earliest days and gave long years of service to the department deserve special mention here; the "redoubtable" Margaret Peake/Cullingford who served as the first Departmental Secretary; for much of her time single-handedly typing all departmental correspondence, teaching materials and research papers, while acting as general administrator and "confidante" to new staff and research students. Secondly, Terry Lymn as Chief Technician, who oversaw the initial move into the Adrian building and then co-ordinated the smooth running of the

Department throughout many years of expansion. Finally, Jenny Foxon, a long time laboratory-based research technician and Trade Union representative who was eventually awarded the MBE for her services to science. They have been succeeded by others who fulfil their roles with the same dedication and professionalism.

Most recently, following departmental reorganisations within the School of Biological Sciences, in 2015 Genetics has welcomed plant geneticists (Prof. David Twell, Prof. Pat-Heslop-Harrison, Dr. Sinead Drea, Dr. Richard Gornall (Director of the University Botanical Gardens), and Dr. James Higgins) and insect social-behavioural biologists (Dr. Eamonn Mallon and Dr. Rob Hammond) from the Department of Biology.

The department now has 33 groups led by internationally renowned scientists and teachers, which embrace all aspects of Genetics across a wide range of organisms including bacteria, fungi, insects, plants, animals and humans. In March 2016 the total staff complement is approximately 200, including PhD students, but excluding MSc and undergraduate students, and Genetics now occupies the whole of the Adrian Building! The growth and continued success of the Department of Genetics, with its many important contributions to science and society, represents the legacy of Bob Pritchard and is a tribute to the insight and inspiration of that remarkable man.

Full details about the current staff complement, and their research and teaching activities, can be found on the Genetics Department web site at: http://www2.le.ac.uk/departments/genetics

About the Authors

Peter Meacock is Emeritus Reader in the Department of Genetics.

Annette Cashmore is a Professor in the Department of Genetics and Director of the Genetics Centre of Excellence in Teaching and Learning, **GENIE***.*

Very sadly and unexpectedly Annette Cashmore passed away on May 28, 2016, shortly after this chapter had been completed. Although she had suffered from the debilitating condition Multiple Sclerosis for several years, she had always continued her roles in teaching and research with great fortitude, passion and enthusiasm right to the very end. She made immense contributions to the success of the Department of Genetics as recorded in this chapter. She was an ardent advocate of synergism between academic teaching and research, collegiality and equality between staff, and the responsibility of scientists to inform and educate society about scientific developments and their implications so that decisions about future research directions could be made on an informed basis; principles that had been instilled by the founder Bob Pritchard. She will be greatly missed.

Year	Notable Milestones in the Department's History
1964	Foundation of the Department of Genetics with Professor Robert Pritchard as Head. Total staff, 7.
1967	The department moves into new accommodation in the Adrian building.
1969	Pritchard, Barth & Collins "Inhibitor Dilution" Model of Bacterial Chromosome Replication published.[1]
1970	Pritchard and Zaritsky publication in *Nature*[2] demonstrates that replication of the bacterial chromosome can be slowed down by limitation of the supply of thymine to an auxotrophic mutant, leading to changes in cellular DNA content.
1975	Opening of the Leicester Medical School which included Genetics teaching as part of the pre-clinical MBChB curriculum. Chandler and Pritchard show that alteration of bacterial chromosome replication by thymine limitation leads to changes in gene dosage consistent with replication proceeding bi-directionally from a unique origin site.[3]
1977	Alec Jeffreys appointed as Lecturer.
1981	First Leicester Cloning Course organised by Prof. Barry Holland.
1982	The Leicester Biocentre, an industry-sponsored research centre, established by Prof. Barry Holland.
1983	Dr. Clive Roberts becomes Head of Department.
1985	Alec Jeffreys publication in *Nature*[4,5] describes the discovery of "DNA-fingerprinting". First immigration dispute resolved using DNA-fingerprinting.[6] Expansion of the Medical School with new Genetics appointments, Charalambos Kyriacou and Raymond Dalgleish.
1986	First criminal investigation using DNA evidence (the Enderby Murders). Alec Jeffreys elected to a Fellowship of the Royal Society.
1987	Prof. Barry Holland becomes Head of Department. MSc Molecular Genetics course launched with Clive Roberts as first Convenor.
1989	25th Anniversary of the Department; "Genetics & Society" public Symposium with invited speakers.
1990	Prof. Peter Williams becomes Head of Department.

(*Continued*)

Table (*Continued*)

Year	Notable Milestones in the Department's History
1992	Prof. Gabriel Dover becomes Head of Department. Alec Jeffreys confirms identification of the mortal remains of Josef Mengele, the "Angel of Death" Nazi war criminal, by DNA-profiling.[7]
1994	Prof. Alec Jeffreys Knighted for services to Genetics. Alec Jeffreys demonstrates that DNA-fingerprinting can be used in Conservation biology to aid breeding programmes for endangered wild species.[8]
1995	Prof. Brian Wilkins becomes Head of Department
1996	BSc in Medical Genetics launched with Dr. Annette Cashmore as Course Director. Prof. Yuri Dubrova, using DNA-fingerprinting techniques on survivors of the 1986 Chernobyl disaster, proves unambiguously that nuclear radiation causes germline mutations in humans.[9] DNA-profiling evidence supports an African origin for modern humans.
1997	All staff of Genetics accommodated within refurbished facilities of the Adrian building.
1998	Prof. Charalambos (Bambos) Kyriacou becomes Head of Department. Alec Jeffreys confirms "Dolly" the sheep to be a true clone by DNA-fingerprinting.[10]
2000	Prof. Annette Cashmore becomes Head of Department. Department of Genetics awarded the highest possible score of 5* in the national UK Research Assessment Exercise.
2002	Queen's Anniversary Prize Awarded to the University for its research in Genetics. Institute of Genetics, a virtual institute to link all genetics researchers irrespective of departments, opened by Prof. Sir Paul Nurse.
2004	40th Anniversary of the Department; research Symposium with talks delivered by current staff members.
2005	Establishment of GENIE CETL with Prof. Annette Cashmore as Director.
2006	Prof. Peter Williams becomes Head of Department, for second time. Prof. Mark Jobling and Dr. Turi King confirm co-ancestry of Y-chromosomes and surnames in British society.[11]
2007	Establishment of the Genomics Core Laboratory. Prof. Mark Jobling and Dr. Turi King demonstrate the genetic legacy of Vikings in the British population.[12] Prof. Mark Jobling and Dr. Turi King demonstrate the Y-chromosome ancestry of Thomas Jefferson, principal author of the American Declaration of Independence, 1776, and Third President of the USA.[13]
2009	Prof. Julian Ketley becomes Head of Department.
2010	Prof. Mark Jobling and Dr. Turi King demonstrate a predominantly Neolithic origin for European paternal ancestry.[14]
2012	Dr. Turi King, working with a multidisciplinary team from the University of Leicester, identifies the mortal remains of the last English Plantagenet King, Richard III.

(*Continued*)

Table (*Continued*)

Year	Notable Milestones in the Department's History
2013	Centre for the study of "Genetic Architecture of Complex Traits" established with Prof. Ed Louis as Director.
2014	Genetics celebrates its 50th Anniversary by a Symposium with talks delivered by current and former departmental researchers. Genetics gains a "100% satisfied" rating for the quality of its teaching in the National Student Survey. Dr. Turi King publishes DNA evidence confirming the identity of the mortal remains of King Richard III.[15]
2015	Public screening of "Code of a Killer", a TV police drama telling how Prof. Alec Jeffreys' DNA fingerprinting technology was used to catch the murderer of two Enderby schoolgirls.
2016	Genetics subsumes staff from the Department of Biology, and occupies the whole of the Adrian building. Total staff ~200.

Key Publications (from Notable Milestones)

1. Pritchard RH, Barth PT, Collins J. (1969) Control of DNA synthesis in bacteria. *Microbial Growth. Symp Soc Gen Microbiol* **19**: 263–297.
2. Pritchard RH, Zaritsky A. (1970) Effect of thymine concentration on the replication velocity of DNA in a thymineless mutant of *Escherichia coli*. *Nature* **226**(5241): 126–131.
3. Chandler MG, Pritchard RH. (1975) The effect of gene concentration and relative gene dosage on gene output in Escherichia coli. *Mol Gen Genet* **138**(2): 127–141.
4. Jeffreys AJ, Wilson V, Thein SL. (1985) Hypervariable 'minisatellite' regions in human DNA. *Nature* **314**(6006): 67–73.
5. Jeffreys AJ, Wilson V, Thein SL. (1985) Individual-specific 'fingerprints' of human DNA. *Nature* **316**(6023): 76–79.
6. Jeffreys AJ, Brookfield JF, Semeonoff R. (1985) Positive identification of an immigration test-case using human DNA fingerprints. *Nature* **317**(6040): 818–819.
7. Jeffreys AJ, Allen MJ, Hagelberg E, Sonnberg A. (1992) Identification of the skeletal remains of Josef Mengele by DNA analysis. *Forensic Sci Int.* **56**(1): 65–76.
8. Signer EN, Schmidt CR, Jeffreys AJ. (1994) DNA variability and parentage testing in captive Waldrapp ibises. *Mol Ecol* **3**(4): 291–300.
9. Dubrova YE, Nesterov VN, Krouchinsky NG, *et al.* (1996) Human minisatellite mutation rate after the Chernobyl accident. *Nature* **380**(6576): 683–686.
10. Signer EN, Dubrova YE, Jeffreys AJ, *et al.* (1998) DNA fingerprinting Dolly. *Nature* **394**(6691): 329–330.

11. King TE, Ballereau SJ, Schürer KE, Jobling MA. (2006) Genetic signatures of coancestry within surnames. *Curr Biol* **16**(4): 384–388.
12. Bowden GR, Balaresque P, King TE, *et al.* (2008) Excavating past population structures by surname-based sampling: the genetic legacy of the Vikings in northwest England. *Mol Biol Evol* **25**(2): 301–309. Epub 2007 Nov 20.
13. King TE, Bowden GR, Balaresque PL, *et al.* (2007) Thomas Jefferson's Y chromosome belongs to a rare European lineage. *Am J Phys Anthropol* **132**(4): 584–589.
14. Balaresque P, Bowden GR, Adams SM, *et al.* (2010) A predominantly Neolithic origin for European paternal lineages. *PLoS Biol* **8**(1): e1000285. doi: 10.1371/journal.pbio.1000285.
15. King TE, Fortes GG, Balaresque P, *et al.* (2014) Identification of the remains of King Richard III. *Nature Communications* **5**, Article number 6631, doi:10.1038/ncomms6631.

Staff of the Department of Genetics from 1964 to 2016

Heads of Department

Prof. Robert Pritchard
Dr. Clive Roberts
Prof. Barry Holland
Prof. Peter Williams
Prof. Gabby Dover
Prof. Brian Wilkins
Prof. Charalambos Kyriacou
Prof. Annette Cashmore
Prof. Julian Ketley

Technical Services Managers

Dr. Susan Hollom (Wilkins)
Terry Lymn
Ila Patel
Dr. Miranda Johnson

Departmental Secretaries

Margaret Peake (Cullingford)
Sandy Davis
Sarah Laband

Other Academic Staff & Independent Research Fellows (alphabetical order)

Dr. Richard Badge
Dr. Julian Barwell
Dr. Chris Bayliss
Dr. Sandra Beleza
Prof. Rhona Borts
Dr. Graham Boulnois
Prof. Tony Brookes
Dr. John Brookfield
Dr. Graeme Bulfield
Prof. Paul Burton (Joint Health Sciences)
Dr. Chris Cane
Dr. Marcus Cooke (Joint Cancer Studies)
Prof. Raymond Dalgleish
Dr. Jennifer Dee
Dr. Sinead Drea
Prof. Yuri Dubrova
Dr. Steven Foster
Dr. Wendy Gibson
Prof. Flaviano Giorgini
Dr. Mark Goodwin
Dr. Richard Gornall
Dr. Bill Grant (RF)
Dr. Rob Hammond
Dr. Bob Hedges
Dr. Mike Hennessey
Prof. Pat Heslop-Harrison
Dr. Colin Hewitt
Dr. James Higgins
Dr. Tim Hirst
Dr. Ed Hollox
Prof. Sir Alec Jeffreys
Prof. Mark Jobling
Dr. Turi King
Dr. Cas Kramer
Prof. Ed Louis
Dr. Eamonn Mallon
Dr. Celia May

Dr. Peter Meacock
Dr. Julie Morrissey
Prof. Marco Oggioni
Dr. Eli Orr
Dr. Mark Plumb
Dr. Catrin Pritchard (RF)
Dr. Ezio Rozato
Dr. Nicola Royle
Dr. Raffael Schaffrath
Dr. Robert Semeonoff
Prof. Nuala Sheehan (Joint Health Sciences)
Dr. Brian Spratt (RF)
Dr. Nicola Suter-Giorgini
Dr. Chris Talbot
Dr. Fred Tata
Dr. Eran Tauber
Prof. Martin Tobin (Joint Health Sciences)
Prof. Richard Trembath
Prof. David Twell

The Department of Genetics, November 2015. Unfortunately the Head of Department, Professor Julian Ketley was busy elsewhere! (Photograph courtesy of Department of Genetics, University of Leicester.)

25 Bob Pritchard — His Passion for Politics

Arnie Gibbons* and Iain Sharpe†

Bob Pritchard came to politics in the 1970s. Prior to that decade British politics had become a sterile affair, completely dominated by two parties (Labour and Conservative), both wedded to "big government" and largely unwilling to rock the boat as long as they got their turn in power. 1974 saw the first tangible signs of a Liberal revival that had been quietly building through the 1960s.

Bob's fundamental belief focused on the relationship between the state and the individual. The state (be it national, local or any other form of authority) exists to enable communities to do things individuals cannot do alone, but the nature of power is such that in doing this the state will inevitably seek to take liberty away from the individual for the convenience of the state. He saw the role of government, politicians and the political process as one of enabling people to achieve their potential and his personal role as one of highlighting and opposing attempts by the state to encroach on individual liberty. The way we do politics in the UK was a constant and unrelenting theme in his contributions to political discourse.

Unimpressed by status, he believed it was the job of Liberals to hold those in power to account. This applied equally to big business as to

*Arnie Gibbons, Leicester, UK (arnie_gibbons2004@yahoo.co.uk).
†Iain Sharpe (iain.sharpe1@ntlworld.com).

insensitive state bureaucracies. He was ever willing to take up issues that he thought right even if they were not electorally popular — for example speaking up for an eccentric constituent he felt had been badly harassed by the police. Drawing on his professional life he often cited DNA fingerprinting as a great advance in shifting the balance from the disbelieving and arbitrary state to the disenfranchised asylum seeker who was now able to prove his or her genetic inheritance.

> *"All Liberals have a profound mistrust of authority and a passionate belief that it is the job of government, and of every Liberal, to defend the individual from the institutions which exercise this authority. We don't seek to undermine these institutions. We want to make them more accountable to us. Accountability strengthens an institution by making it responsive to those who need it and use it." (East Knighton Focus, March 1985)*

Coupled with this belief was Bob's ability to identify and address big issues. Not only did he raise these as theoretical concerns, but many were applied practically at a local level in his day to day conduct of local politics. Furthermore, ideas he vigorously propounded that were at the time seen as "niche" or "wacky" have in many instances become part of mainstream political discourse.

The Liberal Party that Bob joined in Leicester was not in the best of health. In the 1973 local elections it put up candidates in only three out of 16 wards and in 1976 the handful of Liberal standard-bearers finished behind the candidates of the far right National Front with one exception — Knighton ward. Not since 1962 had a Liberal been elected to a council seat in Leicester.

Although Bob's name first appeared on the ballot paper in 1977, it was not until 1982 that his attempts to seek election became serious. In that year he produced the first of what was to become a regular newsletter. By 1999, *Focus* had reached issue number 121. It contained a mix of local and national political issues on which Bob and the Liberal Party (subsequently Liberal Democrats) were campaigning. A typical issue might contain details of a campaign for a pedestrian crossing at a dangerous traffic junction, news of tree planting on a local street, criticism of the City Council for their spending plans on the latest white elephant and an article extolling the work of the European Court of Human Rights and demanding a British Human Rights Act. It gained widespread recognition across the 4,000 or so households who received it and was the key driver of his subsequent electoral success.

At Bob's 70th birthday party in January 2000, some of his scientist friends were teasing him about having abandoned important scientific research for the mundane world of local politics. They picked up a copy of the local newsletter that he issued to his constituents, pointed to a headline and said "This is what you are reduced to worrying about — the location of postboxes". But Bob did not see such concerns as quite so trivial.

Knighton was seen as a solid Conservative fiefdom, being the most affluent part of the city. However, due to its location a significant number of its residents had associations with the University (academics, support staff and students), making it a more liberal community than a superficial glance at the many substantial detached houses in the leafy suburb might suggest.

Through diligent campaigning and his "all year round" presence Bob made inroads into the Conservative majority, first winning election in 1987 and subsequently building a commanding majority.

Notes:

1. *Source: The Elections Centre — Colin Rallings & Michael Thrasher, Plymouth University.*
2. *All elections took place in May and were contested by Liberal (Liberal Democrat from 1989), Conservative and Labour candidates. The Green Party also fielded candidates (1981–84 and 1991 onwards) and the National Front (in 1977 & 1979).*
3. *There were boundary changes prior to the 1983 election.*
4. *Prior to 1997 two tiers of local authority — City and County — were responsible for the functions of local government. From 1997 a single unitary authority (initially elected in 1996 as a "shadow" authority) took over the functions of the two predecessor councils.*

Table 1 Robert Pritchard's Local Government Electoral Record

Year	Ward	Election to	Seats	Votes	%	Result
1977	Knighton	County	2	824	10.9	5th behind two Con and two Lab candidates
1979	Knighton	City	3	1622	15.1	7th behind three Con and three Lab candidates
1981	Castle	County	2	517	12.6	5th behind two Con and two Lab candidates
1983	East Knighton	City	2	675	18.3	Third behind two Con candidates
1984	East Knighton	City	1	858	25.5	Second
1985	East Knighton	County	1	1538	42.7	Second
1987	East Knighton	City	2	2001	46.3	First — elected with one Con
1989	East Knighton	County	1	1968	50.1	First — elected
1991	East Knighton	City	2	1935	49.5	First — elected with second Lib Dem
1993	East Knighton	County	1	2010	59.4	First — elected
1995	East Knighton	City	2	1782	51.5	First — elected with second Lib Dem
1996	East Knighton	Unitary	2	1635	51.4	First — elected with second Lib Dem
1999	East Knighton	Unitary	2	1663	58.4	First — elected with second Lib Dem

Despite fighting less intense election campaigns in later years, Bob's majority continued to rise and was at its highest in 1999 when the Labour runners-up could only manage 19.7%, barely a third of the Lib Dem vote.

Bob also stood for Parliament in 1987, recording 13.9% of the poll in the Leicester South seat (which includes Knighton). This was a tough election for a Liberal candidate to fight as the Conservatives had gained the seat from Labour by just seven votes in the previous election making it the most marginal in the country.

As a campaigner Bob really got into his stride with the Dunlop shareholders campaign. In 1984, before scandals around top people's pay became commonplace, he ran a campaign on behalf of Dunlop shareholders in response to the managing director preaching pay restraint while accepting a big rise himself. ("40,000 small shareholders of Dunlop

Top to bottom: Campaigning on TV; Street Politics; The annual Liberal garden fete, held in Bob's back garden; Winning and Election.

being taken to the cleaners by the banking mafia" in his own words). The campaign achieved considerable national media coverage. As with most of his passions, once elected Bob would seek ways in using his position to further his beliefs. As a member of the Leicestershire County Council Pension Committee he persuaded his colleagues to use the votes attached to the council pension fund shareholdings to oppose unreasonable directors' remuneration packages.

Bob's concern for the environment was the foundation for many of his beliefs and actions. His efforts to promote low energy light bulbs before they were widely available typified this. He not only persuaded the City Council to use them in council buildings and sell them to council tenants, he also sold them mail order through *Focus* to his constituents, spreading the gospel of energy efficiency in the process. Again, long before they became mainstream, he advocated combined heat and power (CHP) projects, and converting council vehicles to run on LPG. He was almost obsessive about getting more trees planted (which he attributed to the influence of his father who had been a botanist) and hugely successful at doing so.

Planning and local government finance are both highly technical areas, but Bob developed a remarkably detailed knowledge of both areas, giving the council professionals a hard time. (However, while he provided a searching challenge to professional officers, he strongly believed that elected politicians should take responsibility for their own decisions and not hide behind the officers.) He believed that planners often did not know best and persistently pressed for more public consultation. The Southern Distributor Road (a ring road planned to cut through the heart of Knighton), the design of which he significantly influenced and threats to the green belt between Knighton and Oadby were the most immediately local of his concerns.

But, as always, he saw the bigger picture. He also opposed ill-thought out greenfield development on the eastern edge of the city, which he felt would lead to poor quality residential neighbourhoods lacking access to services and public transport. At a time when Britain's urban centres faced population exodus and the threat from out of town shopping centres, he was an early advocate for bringing life back to city centres by bringing back residential accommodation — something which is now conventional

Top to bottom: With Lord Mayor; speaking at Leicester City Council; meeting the Queen; Alec Jeffreys getting Freeman of City — Bob with two hats on.

wisdom. Labour City Mayor Sir Peter Soulsby who led the council for most of Bob's twelve years as a councillor said:

> *"We disagreed on a great many things but he was a courteous man and very intelligent. That meant he gave great weight to the opinions of others. He was a passionate advocate of getting more people living in the city centre because of the benefits that would bring. At the time few people agreed with that but over the course of time he has been proved right."*

Because of his wide-ranging grasp of the issues and his ability to take a longer term strategic view he quickly earned the respect of his council group colleagues and served as leader of the Liberal/Lib Dem group on the City Council from 1987 to 1993, a post he relinquished in order to take over the leadership of the Lib Dem group on the County Council, a position he held until local government reorganisation ended his time on the County in 1997.

As a county councillor he provided astute leadership of the Liberal Democrat group at County Hall, in an authority where no one party had overall control. He was a firm believer that getting Lib Dem policies enacted was more important than the trappings of power. By avoiding formal alliances with other parties he ensured that the Lib Dem group flourished. He advocated the view that majority Governments were a bad thing, frequently and passionately. (See his article: *"Why a hung council is a strong council", The Independent, 24 March 1997.*)

He was a strong and consistent champion for local government, opposing centralised "ring fencing", especially when it was proposed by Lib Dem spokespersons who should have known better. He was involved in policy formation at the national level, participating in many policy working groups, but he was never interested in self-promotion for its own sake. At party conferences, Bob was a strong advocate for greater public understanding of science. Perhaps his most prominent, certainly his most frequent, contributions came in defending genetic modification against those who refused to see the benefits of the science in their haste to attack the multinationals who stood to benefit from its exploitation. He regularly contributed to *Liberator*, a radical liberal magazine, with many thought-provoking articles on a wide range of issues reflecting his interests.

On a personal level he was a kind man, who did not mind his house being taken over as a campaign HQ and indeed virtually a Liberal drop-in centre. There were piles of paper everywhere; he would use his stairs as a filing cabinet, piling them up with documents — a different subject on each step. For many years his home hosted the annual "Focus Fete", a fund raiser for the party. Incidentally, the *Focus* leaflets were printed on an offset litho printer housed in Bob's garage (yet another skill he mastered).

It is also worth noting that he was an early pioneer of online communication, using message boards (Prestel) as they were then to discuss things with a handful of other subscribers round the country. He also served as a prison visitor at Gartree prison, and in 1991 jumped out of a plane (see p. 177) in aid of the Lord Mayor's charity — a self-build project for the homeless.

When the Conservatives on Leicester City Council proposed to make footballer Gary Lineker a "freeman of the city" in a fit of blatant populism the Labour administration responded by putting Dr. Alec Jeffreys forward. Most Labour councillors hated Bob as the leader of the Liberal Democrat group on the council and the most effective challenger to the Labour orthodoxy in what was effectively a one-party state. At the ceremony they were mortified when Alec spent part of his speech praising Bob as a professional inspiration.

He was also a political one, inspiring and encouraging those who worked closely with him. There are several Liberal Democrat activists with many years subsequent involvement who would describe him as a major formative political influence. Among those who came into contact with Bob early in their political career was Nick Clegg, who went on to become Deputy Prime Minister (see p. 176). He recognised Bob's contribution in a personal tribute at the time of his passing.

Arnie Gibbons was Liberal Democrat Councillor for East Knighton, Leicester, from 1989–99 and Bob's agent in 1987 when he was first elected.

Iain Sharpe was a Liberal Democrat activist in Leicester in the 1980s and is now a Councillor in Watford. He was Bob's election agent in the 1989 Leicestershire County Council election.

26 Robert Hugh Pritchard: "Bob" to Many — "Dad" to Me

Naomi Matthews (Pritchard)*

This is the true joy of life, the being used for a purpose recognised by yourself as a mighty one; the being a force of nature instead of a feverish, selfish little clod of ailments and grievances complaining that the world will not devote itself to making you happy.

I am of the opinion that my life belongs to the whole community and as long as I live, it is my privilege to do for it whatever I can. I want to be thoroughly used up when I die, for the harder I work the more I live, I rejoice in life for its own sake.

Life is no brief candle to me. It is like a splendid torch which I have got hold of for the moment, and I want to make it burn as brightly as possible before handing it on to future generations.

George Bernard Shaw

If ever a poem encapsulated a person's identity and spirit, then surely this poem could have been written for my dad "Bob".

Robert Hugh Pritchard was born in Wandsworth in 1930 to Florence and Henry Pritchard. He had five older siblings. He was the youngest. He had no real schooling before the age of seven.

* 133 Harrow Road, Wollaton Nottingham, NG8 1FL, UK (naomi@hartzz.com).

Teenager Bob, *ca* 1946.

Bob and his sister Shirley Passmore at Radstock, Somerset, *ca* 1940.

In 1939 when WWII started he was evacuated out of London and sent to Radstock in the coal mining area of Somerset. Although he went there with one of his sisters, Shirley, they were separated and sent to live with strangers under the evacuee scheme.

Fortunately, the family that Bob was sent to recognised that he was very intelligent and arranged for him and his sister to go to the local grammar school.

At the end of the war they returned to London and Bob went to the Emmanuel school in Battersea where he passed his O levels with several distinctions and went on to get his higher school certificate in one year instead of the usual two. He was accepted at King's College London to take a degree in Botany, for which he was awarded a degree with first class honours and the Carter prize for physical biology. From there he went to Glasgow University to study for his PhD as a research student under Guido Pontecorvo.

When, much later, he was interviewed by the "Leicester Mercury", a local newspaper, he was asked who has had the greatest impact on his life he said that it was his mother and Professor Pontecorvo *"who taught me genetics and to be critical"*.

Bob married twice. His first marriage to Jackie produced two sons, John and Simon. When their marriage broke down John and Simon stayed with their father.

Naomi and Bob at Russell Square, London, 1976.

Bob then married Suzi who already had a daughter Naomi — that is where I came into the picture. I think I first met Bob when I was not much more than a baby.

My first memories were staying at a French naturist commune with my mum. Bob came and visited us and my mum used to reminisce that when the gendarme would come looking for Vietnam vets hiding out in the commune Bob would be sent along as the face of respectability to tell them there was no one of that description hiding there!

Dad joked that he may not have given me his genes but he gave me his memes. I'm sure the scientists out there will appreciate that. I know that the other authors in this book will tell you all about what a wonderful scientist he was and how he founded the genetics department at Leicester and so on.

I can only tell you about what is was like to grow up with such a unique man.

Breakfast time was always a forage in the fridge. There were usually more petri dishes in the fridge than food and at the time that was perfectly

Dinner at the Pritchard Family, *ca* 1985.

Left to *right*: Naomi, Simon, Suzi, Bob and John.

normal. When attending a lecture recently about his experiments he was conducting at the time, I was slightly perturbed to discover he was at the time working with botulism amongst other things but we all survived.

As Pontecorvo taught him, so he taught us all, and probably his students too, to be free thinkers. He also taught us to be critical. I can remember when the late Clement Freud came for dinner. He did nothing but complain about the food and the décor and most of the company and nobody said a word. Suddenly as a 13-year-old petulant teenager I took him to task calling him a sexist pig. I was frightened Bob would be angry with me. He told me later he had been very proud of my outburst! Unconventional and unique.

He was the only man I know who used his stairs as a filing cabinet. He said that if he won the lottery the first thing that he would buy would be a secretary. What I enjoyed most about my dad was his incredible wit and wonderfully dry sense of humour.

For example, I remember walking into a restaurant with him and the restaurant was totally empty. The owner asked us if we had a reservation. My father said that we did not and he went off to see if we could be squeezed in. We were laughing so much by the time he came back that I thought he would rather refuse us than have us as his only customers.

My father loved a good debate. Science, politics, religion were all fair topics of interest and nothing was off limits in our household. Although I tried to never ask for help with my homework. He would never be content with helping me to find the right answer. I would be forced to learn the under-pinning theory by which time I had forgotten the answer that I had come to find out in the first place!

One time I came home from school to find Bob chained to a tree outside our house. He stayed there for a long time and did eventually win a reprieve for it. I think the tree will outlast us all.

Bob loved to give people a chance and he always fought for the underdog. Another time I came home from school to find that a prisoner had moved in. He lived with us for quite some time before running off with my birthday money. Bob and Suzi, far from being cross, were concerned as to what had happened to him. He was found safe and well in another prison.

The socializing swimming pool at Knighton Grange Rd, Oadby, *ca* 1988.

The house was generally open to visitors. Many scientists as well as local politicians came to socialise or swim in his swimming pool which was especially popular in the summer.

Some of the Genetics students even put a volley ball net up in the garden and would leave the university after lectures and come for a swim, play a bit of volleyball and maybe discuss some science depending on their moods.

He loved to travel and it was usually combined with Science and a chance to indulge his other passion, which was pinball machines. He always said the book that he would most like to write was "flipping around Europe", which I thought most amusing.

One of my favourite trips was a sabbatical in San Diego California. He rented an apartment right on the boardwalk overlooking the Sea and when not at the University of California he was happy trying the beers in local bars and enjoying the Californian sunshine.

Even then he ended up fighting with the establishment. A man was stabbed close to our apartment. He had no insurance and no hospital wanted to take him. My father was so outraged he remonstrated with them and caused such a fuss that he finally managed to get some treatment for the injured man, maybe saving his life.

I remember once being very excited when he received an invitation to the Queen's garden party at Buckingham Palace. Daughters were allowed to attend as well. Sadly, neither Bob nor Suzi seemed as thrilled as me by the prospect of lunch with the Queen. I said to dad one day — when are we going to that garden party? He fished out the invitation and found out that we had missed it which he found most amusing, although I was somewhat saddened.

Bob's favourite quote was "*It is a strange desire to seek power and to lose liberty*". I think that is why he ran his department much like his home — a bit like a commune!

He despised snobbery and felt that you had to earn respect it could not simply be attributed to you by title.

He was not a great believer in having too many rules or indeed enforcing them. For some it was not a happy regime. For me it taught me to be independent and to be tenacious and proud of it.

Those early debating skills learnt around the family table stood me in good stead for my chosen career in Law. I always thought that if Bob had not been an eminent scientist he would have made a very talented lawyer. In his later role as a councillor he often helped local people who were in trouble with the establishment in one way or another. He would give people jobs, references, and help them to complete forms and applications.

Christmas was always an interesting prospect as you never knew who might turn up. All were welcomed in by Bob and Suzi and would end up being fed even if it was somewhat late due to us all getting distracted. I remember one memorable Christmas when we actually ate our turkey dinner in the early hours of Boxing Day morning as nobody had remembered to put the turkey into the oven earlier that day.

When Bob retired from the University he launched himself into his new career as a Liberal politician with much enthusiasm. University students were replaced with young liberals and a printing press was installed in our kitchen. Pamphlets were churned out at funny times of the day and the night. Everyone was roped in to deliver leaflets and Bob managed to overturn a very safe Conservative seat and held onto it right up until the time he fell ill.

A young Nick Clegg posing with Bob but not sure what they are campaigning about! *ca* 1997/8.

Liberal meetings would be held at our house, budgets would be drafted and campaigns would be plotted. He absolutely adored local politics as it combined so many passions and embraced his desire to help the local community.

Nick Clegg, who was leader of the Liberal Democrat party and deputy prime minister, said about Bob:

> *"Your father was hugely supportive and patient with me when I became the Lib Dem candidate in the East Midlands back in 1997/1998. He was a font of wisdom, good humour and good old fashioned campaigning zeal."*

Yes, that sounds like my dad!

I remember another memorable occasion where his world of politics and Genetics came together. Professor Alec Jeffreys was asked by the

Conservative Councillors to become a free man of Leicester City. They had a huge celebration for this and the Conservatives thought it a major coup. Alec Jeffreys in his speech thanked Bob for giving him the opportunity to do his research at Leicester University and explained that Bob had founded the department. (Bob even turned down a respected job at Oxford to open the department and many thought him crazy to turn that job down for the Leicester opportunity.)

Bob was absolutely delighted. Not because he was publically thanked but because he got one over on his Conservative colleagues. He was such a modest man he had never told anyone about his eminent scientific career and no one had realized the link up to that point.

Another time he telephoned me and said I had better come down to the airfield to watch him as he intended to skydive for charity. I thought him mad and asked why he had agreed to do it. He said he had bet the leader of the Conservatives that he would not do it but they agreed and he could not back out. At nearly 70 years he jumped out of a plane not only

for a good cause but to save face on behalf of the Liberals, an amazing man and I was glad when I watched him land safely.

Sadly, my dad had his share of sorrow and although I like to remember him laughing there were difficult times too. My lovely brother John suffered from schizophrenia and died of an alcohol overdose. They were dark times for our family but somehow made the bond between us stronger, although it was tested at times. Also sadly, the marriage between Suzi my mother and Bob broke down. She moved to Nuneaton to do a ceramics degree and to continue her writing. I remained living with Bob. Many raised an eyebrow that I was still there but neither of us cared. I lived there until I was ready to move on independently.

I remember being worried that he might not approve when I moved out and did it gradually over a 3-months period. In the end I said I was surprised that he had not commented on me moving out. He looked at me and said characteristically that he had not noticed. We both laughed.

After that we led our own lives but we always got together for what Dad described as one of my "working men's dinner" — i.e. it was not the buttered muffins he generally lived on.

My favourite memories are after those meals with a good glass of wine in our hands we would put the world to rights. Politics, Science, world affairs always with a dose of humour but a desperate desire to win our points even if we were playing devil's advocate.

In 2002 Dad was battling Lymphoma and underwent chemotherapy. Unfortunately, the dose was too high and his immune system became compromised. This allowed a virus to enter his brain and cause catastrophic brain damage. We were told that he had about a 4-week life expectancy, but sadly for Bob, he lasted another 12 years. I say sadly because his brain damage was so bad that he lost both his mobility and his power to communicate. When it first happened I thought he would become like Prof. Stephen Hawking and find another way to communicate but sadly he could not even blink his eyes in a yes or no fashion. Worse, he could still laugh and cry and he appeared to comprehend what we were saying.

Those were dark years for both myself and my brother Simon who moved down from London to Leicester to be with his Dad. He visited him and was a constant companion until sadly he developed mouth cancer and passed away three years before his Dad.

It was a strange state of affairs and during the early years occasionally he would miraculously speak. A visitor once came in and was telling one of the caring assistants that he used to be a Professor. Suddenly as if by magic he said "I still am you know".

I used to wonder whether he still had the power but was just too weary to use it under the circumstances. I would often get very frustrated when well-meaning carers would pat him on the head and tell me how sweet he was. I would indignantly tell them that he was not now and had never been sweet natured. Insightful, tenacious, determined, passionate and sometimes ruthless yes, adored by many yes, but never sweet and never docile by choice.

Another sadness for me was that although he got to meet his three lovely grandsons he never got the opportunity to talk to them. I remember they once drew a picture of Grandad Bob on a rollercoaster strapped to his bed. I remember showing them our wedding video and they were amazed that he once had the power to both talk and walk!

I think back to George Bernard Shaw and hope that Bob burned his torch so brightly that he was thoroughly used up by this point and was ready by now to hand it on to the next generation of free and critical thinkers. I see these qualities in droves in his grandchildren brought not ironically by his genes but certainly brought about through the "memes".

A great Scientist yes but so much more..............

With Loving memories

Your Daughter Naomi

Bob and grandsons Matthews, *ca* 2012.

Left to *right*: Shaka Robert (2002; named after Bob), Raphael Benedict (2004), Theo George (2009).

27 Concluding Remarks

Arieh Zaritsky*

This compendium was compiled by a score from Bob's numerous colleagues, former students, political partners and surviving daughter to his memory and inspiration as *"a labour of love"*, using Paul Broda's words (personal communication) about his and Bruce Holloway's tribute to Bill Hayes.[1] Despite various items of criticism, the love, appreciation and gratefulness of each contributor to Bob's memory are distinctly obvious in each chapter. These stem of the optimistic atmosphere that he inspired to all surrounding niches of his presence, during his official "service" and long afterwards — influencing his so-called "intellectual grand-children" through us, his students and colleagues. The freedom-spirit that Bob induced in all of us is reflected in this volume: each composed his/her viewpoint about Bob and affection toward him absolutely independently of the others, hence the variation in style and length among the chapters. The unconditional cooperation mode of Bob is revealed here by the great help received during the editorial work of this book from all authors, to whom I am highly grateful.

In addition to the crucial role of Bob's close family and ancestors, he was in fact an outcome of both his academic mentors, Ponte and Bill, as described in their obituaries ([2,3] and,[1] respectively). To learn much more

*Ben-Gurion University, Be'er-Sheva, Israel (ariehzar@gmail.com).

about Bob and the sources of his scientific, sociological and educational mind, these three memoirs are highly recommended additional reading.

The 25-odd chapters are organized in an almost perfect chronological order, thus telling how various aspects of Bob's extraordinary character might have been evolved and modulated with time. To some extent, there are naturally overlaps, but his variegated, multi-faceted personality is mirroved in each chapter and in the whole volume.

1. Broda P, Holloway B. (1996) William Hayes. 18 January 1913–7 January 1994. *Biogr Mems Fell R Soc* **42**: 172–189.
2. Cohen BL. (2000) Guido Pontecorvo ("Ponte"), 1907–1999. *Genetics* **154**: 497–501.
3. Siddiqi O. (2002) Guido Pontecorvo 29 November 1907–25 September 1999. *Biographical Memoirs of Fellows of the Royal Society* **48**: 375–390. http://www.jstor.org/stable/3560267.

The sources of Bob's personality: Mother, mid-1940s.

Ponte at his microscope, 1975 (the 1950s). (Photograph courtesy of University of Glasgow Archive Services, Guido Pontecorvo collection, GB248 UGC 198/10/3/5/3.)

Bill Hayes, 1980s (the 1960s). (Photograph courtesy of the Australian Academy of Science.)

Index